SECOND QUANTIZATION-BASED METHODS IN QUANTUM CHEMISTRY

SECOND QUANTIZATION-BASED METHODS IN QUANTUM CHEMISTRY

Poul Jørgensen

Chemistry Department
Aarhus University
Aarhus, Denmark

Jack Simons

Chemistry Department
University of Utah
Salt Lake City, Utah

1981

ACADEMIC PRESS

A Subsidiary of Harcourt Brace Jovanovich, Publishers

New York London Toronto Sydney San Francisco

ACADEMIC PRESS, INC.
111 Fifth Avenue, New York, New York 10003

United Kingdom Edition published by
ACADEMIC PRESS, INC. (LONDON) LTD.
24/28 Oval Road, London NW1 7DX

Library of Congress Cataloging in Publication Data

Jørgensen, Poul, Date
Second Quantization-Based Methods in Quantum Chemistry

Includes bibliographies and index.
1. Quantum chemistry. I. Simons, Jack. II. Title.
QD462.J67 541.2'8 81-12880
ISBN 0-12-390220-7 AACR2

PRINTED IN THE UNITED STATES OF AMERICA

81 82 83 84 9 8 7 6 5 4 3 2 1

Contents

Preface

In this book we address several modern quantum chemical tools that are presently being applied at the state-of-the-art level to electronic states of atoms and molecules. We have attempted to concentrate on topics for which textbook coverage does not currently exist in an entirely satisfactory form. The emphasis is on quantum chemical methods whose developments and implementations have been presented in the modern literature primarily in the language of second quantization. We do not assess the precision of the numerical results provided by these methods because many of the techniques discussed are relatively new and their precision limits have not yet been established.

There is little mention of specific molecular systems that might be examined using these tools. We have developed an integrated set of problems with detailed answers, all of which can be worked by hand, to illustrate the practical implementation of the techniques developed. These problems appear at the end of each chapter, and we recommend that they be worked as an integral component of the respective chapters. Excellent treatments of the following very important aspects of quantum chemistry already exist in several texts and are therefore not included in this book: questions of basis set choice, efficient evaluation of requisite one- and two-electron integrals, fast and space-efficient methods for transforming integrals from one basis to another and for storing such integral lists, or the use of orbital symmetry correlation concepts in deciding which electronic configurations must be included for specific molecules. The emphasis here is on describing the structure of the various methods rather than on discussing their numerical implementations.

The choice of topics and depth of presentation were guided by our view of the active research workers who are likely to benefit from this book. Many leading theoretical chemistry research groups have only recently begun to make use of second quantization-based techniques. It is not likely that the full potential of these methods will be realized until those quantum chemists who possess the most sophisticated computational tools and experience become more involved in their use. A presentation that is limited to explaining how working equations of these methods are derived and how the equations are implemented in practice should be especially useful and timely. This monograph is intended to be of use both to the research worker in quantum chemistry and to graduate-level students who have already taken introductory courses that cover the fundamentals of quantum mechanics through the Hartree–Fock method as applied to atoms and molecules. The purpose of this book is more to teach than to survey the literature in the research areas covered. We assume that the reader is familiar with linear algebra, matrix representations of operators, Slater- and contracted Gaussian-type basis functions, the Slater–Condon rules for evaluating determinantal matrix elements of one- and two-electron operators, and the construction of Slater determinant wave functions having proper space–spin symmetry.

List of Abbreviations

BT	Brillouin theorem
BWPT	Brillouin–Wigner perturbation theory
CC	coupled cluster
CHF	coupled Hartree–Fock
CI	configuration interaction
CMCHF	coupled multiconfigurational Hartree–Fock
EOM	equations of motion
EP	electron propagator
GBT	generalized Brillouin theorem
GF	Green's function
GUGA	unitary group, graphical approach
HF	Hartree–Fock
INO	iterative natural orbitals
MBPT	many-body perturbation theory
MCSCF	multiconfigurational self-consistent field
MCTDHF	multiconfigurational time-dependent Hartree–Fock
MPPT	Møller–Plesset perturbation theory
PP	polarization propagator
RHF	restricted Hartree–Fock
RPA	random-phase approximation
RSPT	Rayleigh–Schrödinger perturbation theory
SCF	self-consistent field
SCI	superconfiguration interaction
SP	stationary point
TDHF	time-dependent Hartree–Fock
UHF	unrestricted Hartree–Fock

SECOND QUANTIZATION-BASED METHODS IN QUANTUM CHEMISTRY

Chapter 1 | *Introduction to Second-Quantization Methods*

In the vast majority of the quantum chemistry literature, Slater determinants have been used to express antisymmetric N-electron wavefunctions, and explicit differential and multiplicative operators have been used to write the electronic Hamiltonian. More recently, it has become quite common to express the operators and state vectors that arise in considering stationary electronic states of atoms and molecules (within the Born–Oppenheimer approximation) in the so-called second quantization notation (Linderberg and Öhrn, 1973). The electron creation (r^+, s^+, t^+, u^+) and annihilation (r, s, t, u) operators occurring in this language were originally introduced for use in physical problems that actually involved creation or destruction of particles, photons, or excitations (e.g., phonons). In a majority of the applications of the second-quantization techniques to quantum-chemical problems, no electrons or other particles are created or destroyed. Thus, the operators $\{r^+\}$, $\{r\}$ usually serve merely as a convenient and operationally useful device in terms of which quantum-mechanical states, operators, commutators, and expectation values can be evaluated. In this chapter, we examine how the electronic Hamiltonian, other quantum-mechanical operators, and state vectors are represented in this second-quantization language. We also show how to describe unitary transformations among orthonormal orbitals in an especially convenient manner. In subsequent chapters we make use of the tools of second quantization to describe many approximation techniques (e.g., Hartree–Fock, perturbation theory, configuration interaction, multiconfigurational Hartree–Fock, cluster methods, Green's functions), which are currently in wide use within the quantum chemistry community. The need for such approximation methods is, of course, motivated by our inability to exactly solve electronic structure problems for more than one electron.

A. ANTICOMMUTATION PROPERTIES OF CREATION AND ANNIHILATION OPERATORS

Slater determinantal wavefunctions involving orthonormal spin-orbitals ϕ_k can be represented in terms of products of creation operators on the so-called vacuum ket $|\text{vac}\rangle$,

$$r^+s^+ \cdots t^+|\text{vac}\rangle \leftrightarrow (N!)^{-1/2}\det|\phi_t \cdots \phi_s\phi_r| \equiv |\phi_t \cdots \phi_s\phi_r| \qquad (1.1)$$

The Fermi statistics present in such wavefunctions can be expressed either in terms of a sign change arising upon permuting columns of the determinant or in terms of the following fundamental relation among the r^+ operators:

$$[r^+, s^+]_+ \equiv r^+s^+ + s^+r^+ = 0 \qquad (1.2)$$

Note that this equation also states that the state vector cannot contain the same spin-orbital twice (the Pauli principle) since $r^+r^+ = -r^+r^+ = 0$. Before we go further, we should stress that Eq. (1.1) does not equate Slater determinants to the product of $r^+s^+ \cdots$ operating on $|\text{vac}\rangle$. It simply claims that there is a one-to-one connection between the two objects.

The Fermion annihilation operator r, which is the adjoint of the creation operator r^+, can be thought of as annihilating an electron in ϕ_r and is defined to yield zero when operating on the vacuum ket

$$r|\text{vac}\rangle = 0 \qquad (1.3)$$

The annihilation and creation operators fulfill the following two anticommutation relations (Raimes, 1972):

$$[r, s]_+ \equiv rs + sr = 0 \qquad (1.4)$$

$$[r, s^+]_+ \equiv rs^+ + s^+r = \delta_{rs} \qquad (1.5)$$

which together with Eq. (1.2) comprise the essential relationships used in the application of such second quantization operators to quantum chemistry. For nonorthonormal spin-orbitals, Eq. (1.5) is replaced by $[r, s^+]_+ = \langle\phi_r|\phi_s\rangle$, where the overlap appears explicitly.

The interpretation of Eq. (1.2) in terms of permutational symmetry of determinants is clear. To make the analogous content of Eqs. (1.4) and (1.5) more transparent, we now examine some of the implications that follow from these equations. Let us first examine Eq. (1.5). For $r = s$, this reads $rr^+ + r^+r = 1$. When operating on a ket in which ϕ_r is "occupied," the first term (rr^+) clearly gives zero, since according to Eq. (1.2) terms violating the Pauli principle vanish. The second term (r^+r) yields

$$r^+rt^+u^+ \cdots r^+ \cdots w^+|\text{vac}\rangle = (-1)^{k_r}r^+rr^+t^+u^+ \cdots w^+|\text{vac}\rangle \qquad (1.6)$$

where k_r is the number of creation operators standing to the left of r^+ in the original ket. If this is, according to Eq. (1.5), equal to the original ket, then we must have

$$\begin{aligned} t^+u^+ \cdots r^+ \cdots w^+|\text{vac}\rangle &= (-1)^{k_r} r^+ r r^+ t^+ u^+ \cdots w^+|\text{vac}\rangle \\ &= (-1)^{k_r} r^+ t^+ u^+ \cdots w^+|\text{vac}\rangle \end{aligned} \tag{1.7}$$

The last equality in Eq. (1.7) implies that rr^+, when operating on a ket that does not contain ϕ_r, leaves that ket unchanged, and that r^+r, when acting on a ket in which ϕ_r is present, leaves that ket alone. When r^+r operates on a ket in which ϕ_r is not present, it gives zero. Thus r^+r tells whether orbital ϕ_r occurs in a ket. For that reason, it is often referred to as the occupation number operator $n_r = r^+r$. It is also conventional to introduce the total number operator N as $N = \sum_r n_r$, which when operating on any ket gives as its eigenvalue the total number of electrons in that ket.

In the case $r \neq s$, Eq. (1.5) implies that r operating on any ket that does not contain ϕ_r yields zero, since

$$rs^+t^+u^+ \cdots w^+|\text{vac}\rangle = -s^+rt^+u^+ \cdots w^+|\text{vac}\rangle = 0 \tag{1.8}$$

by repeated use of Eqs. (1.5) and (1.3). When the kets contain both ϕ_r and ϕ_s, both the rs^+ and s^+r terms vanish. For s^+r operating on a ket that contains ϕ_r,

$$\begin{aligned} s^+rt^+u^+ \cdots r^+w^+|\text{vac}\rangle &= (-1)^{k_r} s^+t^+u^+ \cdots w^+|\text{vac}\rangle \\ &= t^+u^+ \cdots s^+w^+|\text{vac}\rangle \end{aligned} \tag{1.9}$$

which is simply a new ket with ϕ_r replaced by ϕ_s.

Finally, we should attempt to elaborate on the meaning of Eq. (1.4). Let us consider the action of rs $(r \neq s)$ on a ket in which ϕ_r and ϕ_s are present:

$$rst^+u^+ \cdots r^+ \cdots s^+ \cdots w^+|\text{vac}\rangle = (-1)^{k_r+k_s} rss^+r^+t^+u^+ \cdots w^+|\text{vac}\rangle$$

which by Eq. (1.5) reduces to

$$(-1)^{k_r+k_s} r(1 - s^+s)r^+t^+u^+ \cdots w^+|\text{vac}\rangle \tag{1.10}$$

The term involving s^+s vanishes because $s|\text{vac}\rangle = 0$, and hence we have (again using $r|\text{vac}\rangle = 0$)

$$(-1)^{k_r+k_s} rr^+t^+u^+ \cdots w^+|\text{vac}\rangle = (-1)^{k_r+k_s} t^+u^+ \cdots w^+|\text{vac}\rangle \tag{1.11}$$

If instead we consider the action of sr, we obtain

$$\begin{aligned} srt^+u^+ \cdots r^+ \cdots s^+ \cdots w^+|\text{vac}\rangle &= (-1)^{k_r+k_s-1} srr^+s^+t^+u^+ \cdots w^+|\text{vac}\rangle \\ &= (-1)^{k_r+k_s-1} t^+u^+ \cdots w^+|\text{vac}\rangle \end{aligned} \tag{1.12}$$

which is opposite in sign to the result of the rs operation. Thus, the statement $rs + sr = 0$ simply means that the effect of annihilation displays Fermion statistics. For $r = s$, Eq. (1.4) reads $rr = -rr = 0$, which also expresses the Pauli principle and Fermi statistics.

Although Eqs. (1.2)–(1.5) contain all of the fundamental properties of the Fermion (electron) creation and annihilation operators, it may be useful to make a few additional remarks about how these operators are used in subsequent applications. In treating perturbative expansions of N-electron wavefunctions or when attempting to optimize the spin-orbitals ϕ_r appearing in such wavefunctions, it is often convenient to refer to Slater determinants that have been obtained from some "reference determinant" by replacing certain spin-orbitals by other spin orbitals. In terms of second-quantized operators, these spin-orbital replacements will be achieved by using the replacement operator s^+r as in Eq. (1.9).

In subsequent chapters, we shall be interested in computing expectation values of one- and two-electron operators. By expressing these operators in terms of the above creation and annihilation operators, the calculation of such expectations values reduces to the evaluation of the elements of the one- and two-electron density matrices $\langle 0|i^+j|0\rangle$ and $\langle 0|i^+j^+lk|0\rangle$ (Davidson, 1976). If the wavefunction $|0\rangle$ is expressed as a linear combination of kets each given in terms of creation operator products $\prod_{r\in 0} r^+|\text{vac}\rangle$, the one- and two-electron density matrices can be evaluated in terms of the expansion coefficients of $|0\rangle$ in these kets. The average occupation of an electron in spin-orbital ϕ_r becomes a particular element $\langle 0|r^+r|0\rangle$ of the one-particle density matrix. If we wish to compute, say, $\langle 0|t^+u|0\rangle$, where $|0\rangle = \prod_{r\in 0} r^+|\text{vac}\rangle$, we may proceed using the anticommutation algebra obeyed by the creation and annihilation operators, to yield

$$\begin{aligned}\langle 0|t^+u|0\rangle &= \prod_{r\in 0}\prod_{r'\in 0}\langle \text{vac}|r't^+ur^+|\text{vac}\rangle \\ &= \langle \text{vac}|r_Nr_{N-1}\cdots r_1t^+ur_1^+r_2^+\cdots r_N^+|\text{vac}\rangle \\ &= \delta_{ur_1}\langle \text{vac}|r_Nr_{N-1}\cdots r_1t^+r_2^+r_3^+\cdots r_N^+|\text{vac}\rangle \\ &\quad -\langle \text{vac}|r_Nr_{N-1}\cdots r_1t^+r_1^+ur_2^+r_3^+\cdots r_N^+|\text{vac}\rangle \qquad (1.13)\end{aligned}$$

which, by "anticommuting" u through to the right (so as to eventually generate $u|\text{vac}\rangle = 0$) and t^+ through to the left (to eventually generate $\langle \text{vac}|t^+ = (t|\text{vac}\rangle)^+ = 0$), and using $\langle \text{vac}|\text{vac}\rangle = 1$, yields a nonvanishing matrix element only when $t = u$ and u is one of the elements of $|0\rangle$ (the "occupied" spin orbitals in $|0\rangle$). This result can be summarized as follows:

$$\langle 0|t^+u|0\rangle = \delta_{tu}v_t \qquad (1.14)$$

where v_t denotes the occupation number of orbital ϕ_t in $|0\rangle$.

This expresses a general rule of how to obtain matrix elements of a replacement operator. The rule is the second-quantization analog of the

Slater–Condon rule (Condon and Shortley, 1935) for evaluating matrix elements of a one-electron operator. In practical calculations one would, of course, use this rule as well as the other counterparts of the Slater–Condon rules.

B. EXPRESSING QUANTUM-MECHANICAL OPERATORS IN SECOND QUANTIZATION

Having now seen how state vectors that are in one-to-one correspondence with N-electron Slater determinants can be represented in terms of Fermion creation and annihilation operators, it still remains for us to show how to express one- and two-electron operators in this language. The second-quantized version of any operator is obtained by simply demanding that the operator, when "sandwiched" between ket vectors of the form $\prod_r r^+|\text{vac}\rangle$, yield exactly the same result as arises in using the first quantized operator between corresponding Slater determinant wavefunctions. For an arbitrary one-electron operator, which in first-quantized language is $\sum_{i=1}^{N} f(\mathbf{r}_i)$, the second quantized equivalent is

$$\sum_{r,s} \langle \phi_r | f | \phi_s \rangle r^+ s \leftrightarrow \sum_{i=1}^{N} f(\mathbf{r}_i) \tag{1.15}$$

where the sums (r, s) are over a complete set of orthonormal spin-orbitals ϕ_r and ϕ_s. The analogous expression for any two-electron operator is

$$\frac{1}{2} \sum_{r,s,t,u} \langle \phi_r \phi_s | g | \phi_t \phi_u \rangle r^+ s^+ u t \leftrightarrow \frac{1}{2} \sum_{i,j=1}^{N} g(\mathbf{r}_i, \mathbf{r}_j) \tag{1.16}$$

Here $\langle \phi_r \phi_s | g | \phi_t \phi_u \rangle$ represents the usual two-electron integral involving the operator g:

$$\langle \phi_r \phi_s | g | \phi_t \phi_u \rangle = \int \phi_r^*(1) \phi_s^*(2) g(1,2) \phi_t(1) \phi_u(2)\, d1\, d2 \tag{1.17}$$

When $g(1,2) = r_{12}^{-1}$, we often express these integrals in short-hand notation as $\langle rs | tu \rangle$. It should be noted that the order of the creation and annihilation operators appearing in Eq. (1.16) must be as presented in order to guarantee that the proper sign will result when expectation and transition value matrix elements of such operators are formed. These spin orbitals $\{\phi_r\}$ are, in most practical applications, obtained as linear combinations of atomic orbital basis functions

$$\phi_r = {}^{\alpha}_{\beta}\sum_a C_{ra} \chi_a \tag{1.18}$$

where α and β are the one-electron spin functions. The χ_a are usually taken to be Slater-type orbitals or contracted Gaussian orbitals, and the C_{ra} are the linear orbital expansion coefficients. In what follows, we assume that the χ_a are real orbitals. This means that the Gaussian or Slater orbitals are given in cartesian form rather than in terms of spherical harmonics.

A few clarifying remarks are now in order. First, one should notice that the first-quantized forms of the above operators contain explicit reference to the number of electrons N, whereas the second-quantized operators do not. This means, for example, that the kinetic energy operator

$$\sum_{r,s} \langle \phi_r | -\tfrac{1}{2}\nabla^2 | \phi_s \rangle r^+ s$$

is independent of N. The kinetic energy operator of the beryllium atom is identical to that of the Be^+, Be^{2+}, Be^-, etc., ions. Of course, nuclear interaction operators $(-Z_A \sum_{i=1}^{N} |\mathbf{r}_i - \mathbf{R}_A|^{-1})$ do contain reference to nuclear charges in their second-quantized version,

$$\sum_{r,s} \langle \phi_r | -Z_A |\mathbf{r} - \mathbf{R}_A|^{-1} | \phi_s \rangle r^+ s$$

but nowhere does N appear. In second quantization, the only reference to N comes from the ket vectors $\prod_{r \in 0} r^+ |\text{vac}\rangle$, which contain N creation operators. This property of operators in the second-quantized language plays an important role, for example, in Green's function methods for calculating ionization energies. The fact that the same Hamiltonian can describe neutral and ion states permits the Green's function to be expressed in terms of a single Hamiltonian.

In examining the above expressions for the second-quantized one- and two-electron operators, it should become clear, for example, that the one-electron operators, which contain $r^+ s$, can "connect" two N-electron kets (corresponding to N-electron Slater determinants) that differ by at most one spin-orbital label. That is, $r^+ s$ can cause only a single spin-orbital replacement. Similarly, the two-electron operators containing $r^+ s^+ ut$ can connect kets differing by at most two spin-orbital labels.

To summarize, we have constructed state vectors that obey Fermi–Dirac statistics through introducing creation and annihilation operators that fulfill the anticommutation relations of Eqs. (1.2), (1.4), and (1.5). The anticommutation relations allow us to build the Slater–Condon rules directly into the operators in the second-quantized language. The operators thereby lose their dependence on the electron number N. The only dependence on N in the second-quantized language appears in the state vectors $|0\rangle$. In contrast, in the first-quantized language the dependence on N appears in both the operators and the wavefunctions.

Because it is important that one fully understand how the above forms of arbitrary one- and two-electron operators are related to the Slater–Condon rules, let us now consider an example of how one uses these operators. Let us compute the expectation value of the electron–electron interaction operator for the three-electron state $2s_\alpha^+ 1s_\beta^+ 1s_\alpha^+ |\text{vac}\rangle$. We know from Slater–Condon type rules that we should get $J_{1s,1s} + 2J_{1s,2s} - K_{1s,2s}$, where J and K are the usual coulomb and exchange integrals:

$$J_{ij} = \langle ij | ij \rangle \tag{1.19}$$

$$K_{ij} = \langle ij | ji \rangle \tag{1.20}$$

The second-quantized approach involves evaluating

$$\frac{1}{2} \sum_{r,s,t,u} \langle \text{vac} | 1s_\alpha 1s_\beta 2s_\alpha r^+ s^+ u t 2s_\alpha^+ 1s_\beta^+ 1s_\alpha^+ | \text{vac} \rangle \langle \phi_r \phi_s | \frac{1}{r_{12}} | \phi_t \phi_u \rangle$$

which involves the two-particle density matrix.

The application of Eq. (1.5) in the form $rs^+ = \delta_{rs} - s^+ r$ permits the annihilation operators u, t to be anticommuted to the right in the above and the creation operators $r^+ s^+$ to be moved to the left. This strategy permits us to identify all of the nonvanishing contributions (those arising from the δ_{rs} terms) and to eventually obtain $u|\text{vac}\rangle$ *or* $\langle \text{vac}| r^+$, both of which yield zero. The process of moving ut to the right is carried out as follows:

$$\begin{aligned}
ut2s_\alpha^+ 1s_\beta^+ 1s_\alpha^+ |\text{vac}\rangle &= u(\delta_{t2s_\alpha} - 2s_\alpha^+ t) 1s_\beta^+ 1s_\alpha^+ |\text{vac}\rangle \\
&= [\delta_{t2s_\alpha}(\delta_{u1s_\beta} - 1s_\beta^+ u) 1s_\alpha^+ - u2s_\alpha^+ (\delta_{t1s_\beta} - 1s_\beta^+ t) 1s_\alpha^+] |\text{vac}\rangle \\
&= [\delta_{t2s_\alpha}\delta_{u1s_\beta} 1s_\alpha^+ - \delta_{t2s_\alpha}\delta_{u1s_\alpha} 1s_\beta^+ \\
&\quad - (\delta_{u2s_\alpha} - 2s_\alpha^+ u)(\delta_{t1s_\beta} - 1s_\beta^+ t) 1s_\alpha^+] |\text{vac}\rangle \\
&= [\delta_{t2s_\alpha}\delta_{u1s_\beta} 1s_\alpha^+ - \delta_{t2s_\alpha}\delta_{u1s_\alpha} 1s_\beta^+ - \delta_{u2s_\alpha}\delta_{t1s_\beta} 1s_\alpha^+ \\
&\quad + \delta_{u2s_\alpha}\delta_{t1s_\alpha} 1s_\beta^+ + \delta_{t1s_\beta}\delta_{u1s_\alpha} 2s_\alpha^+ - \delta_{t1s_\alpha}\delta_{u1s_\beta} 2s_\alpha^+] |\text{vac}\rangle
\end{aligned} \tag{1.21}$$

The treatment of $\langle \text{vac} | 1s_\alpha 1s_\beta 2s_\alpha r^+ s^+$ goes through in exactly the same manner and yields the adjoint of the above result, with r replacing t and s replacing u:

$$\begin{aligned}
\langle \text{vac} | 1s_\alpha 1s_\beta 2s_\alpha r^+ s^+ = \langle \text{vac} | [&\delta_{r2s_\alpha}\delta_{s1s_\beta} 1s_\alpha - \delta_{r2s_\alpha}\delta_{s1s_\alpha} 1s_\beta - \delta_{s2s_\alpha}\delta_{r1s_\beta} 1s_\alpha \\
&+ \delta_{s2s_\alpha}\delta_{r1s_\alpha} 1s_\beta + \delta_{r1s_\beta}\delta_{s1s_\alpha} 2s_\alpha - \delta_{r1s_\alpha}\delta_{s1s_\beta} 2s_\alpha]
\end{aligned} \tag{1.22}$$

Then by forming the scalar product $\langle \text{vac} | 1s_\alpha 1s_\beta 2s_\alpha r^+ s^+ u t 2s_\alpha^+ 1s_\beta^+ 1s_\alpha^+ | \text{vac} \rangle$ and using, for example, the fact that $\langle \text{vac} | 1s_\alpha 1s_\alpha^+ | \text{vac} \rangle = 1$ and $\langle \text{vac} | 1s_\alpha 2s_\alpha^+ | \text{vac} \rangle = 0$, one obtains

$$\begin{aligned}
&\delta_{tr2s_\alpha}\delta_{us1s_\beta} + \delta_{tr2s_\alpha}\delta_{su1s_\alpha} + \delta_{su2s_\alpha}\delta_{tr1s_\beta} + \delta_{su2s_\alpha}\delta_{tr1s_\alpha} \\
&\quad + \delta_{tr1s_\beta}\delta_{su1s_\alpha} + \delta_{tr1s_\alpha}\delta_{su1s_\beta} - \delta_{ur2s_\alpha}\delta_{ts1s_\beta} - \delta_{ur1s_\beta}\delta_{ts2s_\alpha} \\
&\quad - \delta_{ts2s_\alpha}\delta_{ur1s_\alpha} - \delta_{ur2s_\alpha}\delta_{ts1s_\alpha} - \delta_{ts1s_\alpha}\delta_{ru1s_\beta} - \delta_{ts1s_\beta}\delta_{ru1s_\alpha}
\end{aligned}$$

where the triple-indexed delta function δ_{ijk} means $\delta_{ij}\delta_{jk}$. This result, when multiplied by $\frac{1}{2}\langle\phi_r\phi_s|1/r_{12}|\phi_t\phi_u\rangle$ and summed over r, s, t, u, indeed yields

$$J_{1s,1s} + 2J_{1s,2s} - K_{1s,2s}$$

C. TENSOR OPERATORS

To gain further experience and understanding, let us also consider how specific operators that are familiar in first quantization-notation are mapped into their second-quantized analogs. The z projection of the total spin S_z is given by

$$S_z = \sum_{r,s} \langle\phi_r|S_z|\phi_s\rangle r^+ s \tag{1.23}$$

which, if the m_s dependence of the spin-orbitals is made explicit ($\phi_r = {}^{\alpha}_{\beta}\bar{\phi}_r$), becomes

$$S_z = \tfrac{1}{2}\hbar \sum_{r,s} \langle\bar{\phi}_r|\bar{\phi}_s\rangle [r_\alpha^+ s_\alpha - r_\beta^+ s_\beta] \tag{1.24}$$

Because the orbitals $\bar{\phi}_r$ are assumed to be orthonormal, spatial integration further reduces this to

$$S_z = \sum_r [r_\alpha^+ r_\alpha - r_\beta^+ r_\beta]\tfrac{1}{2}\hbar \tag{1.25}$$

where the sum is over the orbitals $\bar{\phi}_r$. The spin-raising and -lowering operators $S_\pm = S_x \pm iS_y$ are, in second quantized form,

$$S_+ = \sum_{r,s} \langle\phi_r|S_+|\phi_s\rangle r^+ s = \sum_r \hbar r_\alpha^+ r_\beta, \tag{1.26}$$

$$S_- = \sum_r \hbar r_\beta^+ r_\alpha \tag{1.27}$$

In addition to the operators discussed above, it is often important in quantum-chemical applications to evaluate commutators of pairs of operators. For example, to show that the creation operator r_α^+ is of doublet spin character (i.e., has the potential to change the total spin eigenvalue of any function upon which it acts by $\pm\frac{1}{2}\hbar$) it is sufficient to demonstrate that $[S_z, r_\alpha^+] = \frac{1}{2}\hbar r_\alpha^+$, $[S_-, r_\alpha^+] = \hbar r_\beta^+$, and $[S_+, r_\alpha^+] = 0$. As an example of how to evaluate such commutators, let us compute $[S_z, r_\alpha^+]$ and $[S_-, r_\alpha^+]$:

$$[S_z, r_\alpha^+] = \frac{\hbar}{2} \sum_t [t_\alpha^+ t_\alpha - t_\beta^+ t_\beta, r_\alpha^+] \tag{1.28}$$

Now

$$t_\alpha^+ t_\alpha r_\alpha^+ - r_\alpha^+ t_\alpha^+ t_\alpha = \delta_{tr} t_\alpha^+ - t_\alpha^+ r_\alpha^+ t_\alpha - r_\alpha^+ t_\alpha^+ t_\alpha = \delta_{tr} t_\alpha^+ \tag{1.29}$$

and also

$$t_\beta^+ t_\beta r_\alpha^+ - r_\alpha^+ t_\beta^+ t_\beta = -t_\beta^+ r_\alpha^+ t_\beta - r_\alpha^+ t_\beta^+ t_\beta = 0 \tag{1.30}$$

Therefore,

$$[S_z, r_\alpha^+] = \tfrac{1}{2}\hbar r_\alpha^+ \tag{1.31}$$

Before moving on, we wish to clearly point out an important consequence of performing commutation between operator pairs. For example, notice that although each term in the commutator arising in Eq. (1.29) involves three operators (e.g., $t_\alpha^+ t_\alpha r_\alpha^+$), the final result contains only one operator. This reduction of the operator "rank" always arises when performing such commutators. We usually say that an operator such as $r^+ s$ has one-particle rank, whereas $r^+ s^+ tu$ has two-particle rank. Such rank lowering is an important feature, which is explicitly brought about in the second-quantized language and which is used on numerous occasions in Chapter 2. Because the second-quantized operators contain no reference to N, this cancellation can be achieved at the operator level. The same cancellation occurs in first-quantization calculations but not until determinantal matrix elements are taken.

The commutator involving S_- can be written as

$$\begin{aligned}
[S_-, r_\alpha^+] &= \hbar \sum_t [t_\beta^+ t_\alpha, r_\alpha^+] = \hbar \sum_t (t_\beta^+ t_\alpha r_\alpha^+ - r_\alpha^+ t_\beta^+ t_\alpha) \\
&= \hbar \sum_t (\delta_{rt} t_\beta^+ - t_\beta^+ r_\alpha^+ t_\alpha - r_\alpha^+ t_\beta^+ t_\alpha) \\
&= \hbar r_\beta^+
\end{aligned} \tag{1.32}$$

(Again, note the reduction in particle rank.) The importance of this result is that r^+ when operating on any eigenfunction of S^2 (e.g., $S^2|0\rangle = \hbar s(s+1)|0\rangle$) will yield a function whose S_z eigenvalue $m_s\hbar$ is increased by $\frac{1}{2}\hbar$

$$S_z r_\alpha^+|0\rangle = r_\alpha^+ S_z|0\rangle + \tfrac{1}{2}\hbar r_\alpha^+|0\rangle = (m_s + \tfrac{1}{2})\hbar r_\alpha^+|0\rangle$$

As defined in group theory (Tinkham, 1964), general tensor operators of rank L obey $[J_z, T_\mu^L] = \mu\hbar T_\mu^L$ and $[J_\pm, T_\mu^L] = \hbar[L(L+1) - \mu(\mu \pm 1)]^{1/2} T_{\mu\pm 1}^L$, where $\mathbf{J}$ refers to angular momentum. Our operators r_α^+, r_β^+ correspond to $L = \frac{1}{2}$, $\mu = \pm\frac{1}{2}$. These operators, together with their corresponding annihilation partners ($r_\alpha = T_{-1/2}^{1/2}$, $r_\beta = -T_{1/2}^{1/2}$), can then be combined, using vector-coupling coefficients to generate two-or-more-electron creation operators having various total spin values. For example, the two-electron pair creation singlet tensor operator is

$$T_{rs}^+(0,0) = \frac{1}{\sqrt{2}}[r_\alpha^+ s_\beta^+ - r_\beta^+ s_\alpha^+] \tag{1.33}$$

whereas the two-electron triplet (with various M_s) tensor operators are

$$\begin{aligned} T_{rs}^{+}(1,1) &= r_{\alpha}^{+}s_{\alpha}^{+} \\ T_{rs}^{+}(1,-1) &= r_{\beta}^{+}s_{\beta}^{+} \\ T_{rs}^{+}(1,0) &= \frac{1}{\sqrt{2}}\left[r_{\alpha}^{+}s_{\beta}^{+} + r_{\beta}^{+}s_{\alpha}^{+}\right] \end{aligned} \tag{1.34}$$

The operators that create singlet and triplet coupled one-electron replacements are

$$\begin{aligned} &q_{rs}^{+}(0,0) = \frac{1}{\sqrt{2}}(r_{\alpha}^{+}s_{\alpha} + r_{\beta}^{+}s_{\beta}), \qquad q_{rs}^{+}(1,1) = -r_{\alpha}^{+}s_{\beta} \\ &q_{rs}^{+}(1,-1) = r_{\beta}^{+}s_{\alpha}, \qquad q_{rs}^{+}(1,0) = \frac{1}{\sqrt{2}}\left[r_{\alpha}^{+}s_{\alpha} - r_{\beta}^{+}s_{\beta}\right] \end{aligned} \tag{1.35}$$

Such tensor operators often occur when one- and two-electron operators are expressed in the second-quantization language. For example, the electronic part of the electric dipole operator $e\mathscr{E}\cdot\sum_{i=1}^{N}\mathbf{r}_i$ becomes $\sum_{t,s} e\mathscr{E}\cdot\langle\phi_t|\mathbf{r}|\phi_s\rangle\, t^{+}s$, which after spin integration reduces to

$$\sum_{t,s} e\mathscr{E}\cdot\langle\Phi_t|\mathbf{r}|\Phi_s\rangle\left[t_{\alpha}^{+}s_{\alpha} + t_{\beta}^{+}s_{\beta}\right]$$

which contains the singlet-spin tensor operator.

D. UNITARY TRANSFORMATIONS OF ORBITALS

Having now been introduced to the basic properties of Fermion creation and annihilation operators as they express N-electron wavefunctions and quantum-mechanical operators, as well as to the strategy involved in manipulating these operators, we are nearly ready to consider the efficient use of these tools in expressing wavefunctions as they are actually employed in state-of-the-art quantum-chemical studies. It frequently occurs that we are in possession of a set of orthonormal spin orbitals that, although their construction was straightforwardly achieved, may not represent an optimal choice for the problem under consideration. Hence, it is natural to consider how one can describe unitary transformations among these orbitals within the second-quantization language. We have already shown that the replacement operator $r^{+}s$ yields, when operating on a ket in which ϕ_s is occupied, a new ket with ϕ_s replaced by ϕ_r. Now we wish to demonstrate that the exponential operator $\exp(i\lambda)$ defined as

$$\exp(i\lambda) = 1 + i\lambda + \frac{1}{2!}(i\lambda)(i\lambda) + \cdots \tag{1.36}$$

where

$$\lambda = \sum_{r,s} \lambda_{rs} r^+ s \tag{1.37}$$

yields, when operating on *any* ket (and hence when operating on *any* wavefunction that is expressed as a linear combination of kets), a new ket in which each spin-orbital of the original ket (ϕ_t) is transformed into a new spin-orbital $\tilde{\phi}_t$ (Dalgaard and Jørgensen, 1978). Let us therefore consider the effect of such an exponential transformation $[\exp(i\lambda)]$ on an arbitrary ket. Using the fact that $[\exp(i\lambda)]^{-1} = \exp(-i\lambda)$, we may show the following:

$$\begin{aligned} \exp(i\lambda)t_1^+ t_2^+ \cdots t_N^+|\text{vac}\rangle &= \exp(i\lambda)t_1^+ \exp(-i\lambda)\exp(i\lambda)t_2^+ \exp(-i\lambda) \\ &\quad \cdots \exp(i\lambda)t_N^+ \exp(-i\lambda)\exp(i\lambda)|\text{vac}\rangle \end{aligned} \tag{1.38}$$

which because $\lambda|\text{vac}\rangle = 0$ becomes

$$\tilde{t}_1^+ \tilde{t}_2^+ \cdots \tilde{t}_N^+ |\text{vac}\rangle \tag{1.39}$$

with the modified creation operators being defined by

$$\tilde{t}_k^+ \equiv \exp(i\lambda)t_k^+ \exp(-i\lambda) \tag{1.40}$$

By now expanding both of the above exponential operators, we obtain

$$\tilde{t}_k^+ = t_k^+ + i[\lambda, t_k^+] + \frac{i^2}{2!}[\lambda,[\lambda,t_k^+]] + \cdots \tag{1.41}$$

Because

$$[\lambda, t_k^+] = \sum_{r,s} \lambda_{rs}[r^+ s, t_k^+] = \sum_r \lambda_{rt_k} r^+ \tag{1.42}$$

$$[\lambda,[\lambda,t_k^+]] = \sum_{r,s} \lambda_{sr}\lambda_{rt_k} s^+ = \sum_s (\lambda\lambda)_{st_k} s^+ \tag{1.43}$$

Eq. (1.41) can be rewritten

$$\begin{aligned} \tilde{t}_k^+ &= t_k^+ + i \sum_s \lambda_{st_k} s^+ + \frac{i^2}{2!} \sum_s (\boldsymbol{\lambda\lambda})_{st_k} s^+ \\ &\quad + \frac{i^3}{3!} \sum_s (\boldsymbol{\lambda\lambda\lambda})_{st_k} s^+ + \cdots = \sum_s [\exp(i\boldsymbol{\lambda})]_{st_k} s^+ \end{aligned} \tag{1.44}$$

The exponential matrix $\exp(i\boldsymbol{\lambda})$ appearing in Eq. (1.44) is defined through the power series appearing in that equation. However, as we show below, this matrix can be computed from the $\boldsymbol{\lambda}$ matrix in a much more straightforward and practical manner.

If we want the transformation described by $\exp(i\lambda)$ to preserve orthonormality of the spin-orbitals or, equivalently, to preserve the anticommutation relations [see discussion following Eq. (1.5)]

$$[t^+, s]_+ = [\tilde{t}^+, \tilde{s}]_+ = \delta_{ts} \tag{1.45}$$

then certain restrictions must be placed on λ. This restriction, of course, requires the above transformation to be unitary. Using Eq. (1.40) and

$$[\exp(i\lambda)]^{+} = \exp(-i\lambda^{+}) \tag{1.46}$$

then

$$\tilde{s} = \exp(i\lambda^{+})s\exp(-i\lambda^{+}) \tag{1.47}$$

We therefore have

$$\begin{aligned}\tilde{t}^{+}\tilde{s} + \tilde{s}\tilde{t}^{+} &= \exp(i\lambda)t^{+}\exp(-i\lambda)\exp(i\lambda^{+})s\exp(-i\lambda^{+}) \\ &\quad + \exp(i\lambda^{+})s\exp(-i\lambda^{+})\exp(i\lambda)t^{+}\exp(-i\lambda)\end{aligned} \tag{1.48}$$

Now if the operator λ is required to be hermitian, which then makes the elements λ_{rs} form a hermitian matrix

$$\lambda^{+} = \sum_{r,s}(\lambda_{rs}r^{+}s)^{+} = \sum_{r,s}\lambda_{rs}^{*}s^{+}r = \sum_{r,s}\lambda_{sr}s^{+}r = \lambda \tag{1.49}$$

then Eq. (1.48) will reduce to

$$\tilde{t}^{+}\tilde{s} + \tilde{s}\tilde{t}^{+} = \exp(i\lambda)(t^{+}s + st^{+})\exp(-i\lambda) = \delta_{ts} \tag{1.50}$$

which means that the above transformation does indeed preserve the anti-commutation relations. The fact that the $\boldsymbol{\lambda}$ matrix is hermitian implies that the transformation matrix $\exp(i\boldsymbol{\lambda})$ occurring in Eq. (1.44) is unitary since

$$[\exp(i\boldsymbol{\lambda})]^{+} = \exp(-i\boldsymbol{\lambda}^{+}) = \exp(-i\boldsymbol{\lambda}) = [\exp(i\boldsymbol{\lambda})]^{-1} \tag{1.51}$$

This means that the orbital transformation

$$\tilde{\phi}_r \equiv \sum_{s}[\exp(i\boldsymbol{\lambda})]_{sr}\phi_s \tag{1.52}$$

is also unitary and hence preserves orthonomality.

When the matrix $\boldsymbol{\lambda}$ is hermitian, it can be divided into real and imaginary parts

$$\boldsymbol{\lambda} = \boldsymbol{\alpha} + i\boldsymbol{\kappa} \tag{1.53}$$

where the matrix $\boldsymbol{\alpha}$ is real and symmetric ($\alpha_{rs} = \alpha_{sr}$) and the matrix $\boldsymbol{\kappa}$ is real and antisymmetric ($\kappa_{rs} = -\kappa_{sr}$, i.e., $\kappa_{rr} = 0$). The operator λ may also be divided,

$$\begin{aligned}\lambda &= \sum_{rs}\lambda_{rs}r^{+}s = \sum_{rs}(\alpha_{rs} + i\kappa_{rs})r^{+}s \\ &= \sum_{r}\alpha_{rr}r^{+}r + \sum_{r>s}\alpha_{rs}(r^{+}s + s^{+}r) + i\sum_{r>s}\kappa_{rs}(r^{+}s - s^{+}r)\end{aligned} \tag{1.54}$$

into three terms each of which are hermitian. Since we have assumed earlier that the spin-orbitals are real, orbital variations in $\exp(i\lambda)$ described through the α_{rs} parameters must vanish identically because these variations would map the real orbitals into complex orbitals. When the variations described by the α_{rs} parameters are eliminated, the unitary transformation described by $\exp(i\boldsymbol{\lambda})$ becomes an orthogonal transformation $\exp(-\boldsymbol{\kappa})$. In what follows, when we refer to a unitary transformation, it is usually the orthogonal transformation described above.

To see how one can express the $\exp(i\boldsymbol{\lambda})$ matrix most compactly, let us introduce the unitary transformation u, which diagonalizes $\boldsymbol{\lambda}$:

$$\mathsf{udu}^+ = \boldsymbol{\lambda}, \quad \mathsf{uu}^+ = \mathsf{u}^+\mathsf{u} = 1, \qquad d_{rs} = \delta_{rs} d_r \tag{1.55}$$

Then

$$\begin{aligned}
\exp(i\boldsymbol{\lambda}) &= 1 + i\boldsymbol{\lambda} + \frac{i^2}{2!}\boldsymbol{\lambda\lambda} + \cdots \\
&= \mathsf{uu}^+ + i\mathsf{udu}^+ + \frac{i^2}{2!}\mathsf{udu}^+\mathsf{udu}^+ + \cdots \\
&= \mathsf{u}\left(1 + i\mathsf{d} + \frac{i^2}{2!}\mathsf{dd} + \frac{i^3}{3!}\mathsf{ddd} + \cdots\right)\mathsf{u}^+ \\
&= \mathsf{u}\exp(i\mathsf{d})\mathsf{u}^+
\end{aligned} \tag{1.56}$$

Because d is diagonal $(\mathsf{dd})_{rs} = \delta_{rs} d_r^2$ and hence $[\exp(i\mathsf{d})]_{rs} = \delta_{rs}\exp(id_r)$. Therefore the elements of $\exp(i\boldsymbol{\lambda})$ are easily given by

$$[\exp(i\boldsymbol{\lambda})]_{rs} = \sum_t u_{rt}\exp(id_t)u_{ts}^+ \tag{1.57}$$

This equation gives a compact and efficient expression for the orbital transformation matrix appearing in Eq. (1.52).

Having seen how the operators of second quantization can be used to express wavefunctions and quantum-mechanical operators, let us now move on to the problem of choosing wavefunctions that yield optimum descriptions, in an energy optimization sense, of the stationary states of atomic and molecular systems.

PROBLEMS

1.1 Show the following identities to be valid for the operators A, B, and C:

$$[AB, C] = A[B, C] + [A, C]B = A[B, C]_+ - [A, C]_+ B$$

$$[AB, C]_+ = [A, C]_+ B + A[B, C]$$

$$[A_1 A_2 \cdots A_N, B] = \sum_{j=1}^{N} A_1 A_2 \cdots A_{j-1}[A_j, B]A_{j+1} \cdots A_N$$

1.2

1. Show by performing the following commutators and anticommutators that one achieves a lowering of the particle rank of the operators involved:

$$[i^+j, k^+l], \qquad [s, [i^+j^+lk, r^+]]_+, \qquad [r^+s, [i^+j, t^+v]]$$

2. By how much (what order) has the rank been lowered in each case?

1.3 Show that the operators $\{q_r{}^+{}_s(1, i);\ i = 1, 0, -1\}$ given in Eq. (1.35) are tensor operators in spin space with $S = 1$.

1.4 The Fermi contact Hamiltonian may, in first-quantized language, be written as

$$H_F = \sum_i^N \sum_a \frac{8\pi}{3} g \cdot \beta \cdot \gamma_a \mathbf{S}_i \cdot \mathbf{I}^a \delta(\mathbf{r}_i - R_a)$$

where $\mathbf{I}^a$ is the spin of nucleus a, and g is the electronic gyromagnetic ratio, β the Bohr magneton, γ_a the nuclear gyromagnetic factor, $\mathbf{S}$ the electron spin, and $\delta(\mathbf{r})$ the Dirac delta function. Show that this Hamiltonian, in second-quantized language, may be written as

$$H_F = \sum_{a,r,s} \frac{4\pi}{3} g\beta\gamma_a \bar{\phi}_r^*(R_a)\bar{\phi}_s(R_a)[-q_{rs}^+(1,1)(I_x^a - iI_y^a)$$

$$+ q_{rs}^+(1,-1)(I_x^a + iI_y^a) + \sqrt{2}q_{rs}^+(1,0)I_z^a]$$

where $\bar{\phi}_s(R_a)$ is the amplitude of orbital s at nucleus a and the q^+ operators are defined in Eq. (1.35).

1.5 Given two orthonormal orbitals ϕ_1 and ϕ_2 expressed as linear combinations of two not necessarily orthonormal basis functions χ_a and χ_b:

$$\phi_{1,2} = a_{1,2}\chi_a + b_{1,2}\chi_b$$

1. Show that the two-dimensional unitary transformation given in Eq. (1.52), which in this case is described by a $\boldsymbol{\kappa}$ matrix

$$\boldsymbol{\kappa} = \begin{pmatrix} 0 & -\kappa \\ \kappa & 0 \end{pmatrix}$$

involving one parameter κ, can be expressed as

$$\exp(-\boldsymbol{\kappa}) = \cos\kappa \begin{pmatrix} 1 & 0 \\ 0 & 1 \end{pmatrix} + \sin\kappa \begin{pmatrix} 0 & 1 \\ -1 & 0 \end{pmatrix} = \begin{pmatrix} \cos\kappa & \sin\kappa \\ -\sin\kappa & \cos\kappa \end{pmatrix}$$

2. Apply this transformation to the orbitals $\phi_{1,2}$ for $\kappa = 10°$ to obtain new orbitals $\tilde{\phi}_{1,2}$ and express $\tilde{\phi}_{1,2}$ explicitly in terms of $\chi_a, \chi_b, a_{1,2}, b_{1,2}$.

SOLUTIONS

1.1

$$\begin{aligned}[AB, C] &= ABC - CAB = ABC - ACB - CAB + ACB \\ &= A[B, C] + [A, C]B = A[B, C]_+ - [A, C]_+B \\ [AB, C]_+ &= ABC + CAB = ABC - ACB + ACB + CAB \\ &= [A, C]_+B + A[B, C]\end{aligned}$$

$$\begin{aligned}[A_1A_2 \cdots A_N, B] &= A_1A_2 \cdots A_NB - BA_1A_2 \cdots A_N \\ &= A_1A_2 \cdots A_NB - A_1A_2 \cdots BA_N \\ &\quad + A_1A_2 \cdots BA_N - BA_1A_2 \cdots A_N \\ &= A_1A_2 \cdots [A_N, B] + A_1A_2 \cdots [A_{N-1}, B]A_N \\ &\quad + A_1A_2 \cdots A_{N-2}BA_{N-1}A_N - BA_1A_2 \cdots A_N\end{aligned}$$

etc. Clearly, by continuing to move the B to the left, we generate all terms in the series:

$$\sum_{j=1}^{N} A_1 \cdots A_{j-1}[A_j, B]A_{j+1} \cdots A_N$$

1.2

1. $$\begin{aligned}[i^+j, k^+l] &= i^+jk^+l - k^+li^+j \\ &= \delta_{jk}i^+l - i^+k^+jl - \delta_{li}k^+j + k^+i^+lj \\ &= \delta_{jk}i^+l - \delta_{li}k^+j \\ [i^+j^+lk, r^+] &= \delta_{kr}i^+j^+l - \delta_{lr}i^+j^+k\end{aligned}$$

Hence

$$\begin{aligned}[s, [i^+j^+lk, r^+]]_+ &= \delta_{kr}(si^+j^+l + i^+j^+ls) - \delta_{lr}(si^+j^+k + i^+j^+ks) \\ &= \delta_{kr}(\delta_{is}j^+l - \delta_{js}i^+l) - \delta_{lr}(\delta_{is}j^+k - \delta_{js}i^+k) \\ [i^+j, t^+v] &= \delta_{jt}i^+v - \delta_{iv}t^+j\end{aligned}$$

Hence

$$\begin{aligned}[r^+s, [i^+j, t^+v]] &= \delta_{jt}[r^+s, i^+v] - \delta_{iv}[r^+s, t^+j] \\ &= \delta_{jt}(\delta_{si}r^+v - \delta_{rv}i^+s) - \delta_{iv}(\delta_{ts}r^+j - \delta_{rj}t^+s)\end{aligned}$$

2. In the first problem we lowered the rank by one, in the second by two, and in the third by two.

1.3

$$\{q_{rs}^+(1, M_s)\} = \left\{-r_\alpha^+ s_\beta, \frac{1}{\sqrt{2}}(r_\alpha^+ s_\alpha - r_\beta^+ s_\beta), r_\beta^+ s_\alpha\right\}$$

Consider the $M_s = 0$ component only (the others are treated in like fashion):

$$[S_z, q_{rs}^+(1,0)] = \sum_j \frac{1}{\sqrt{2}} [j_\alpha^+ j_\alpha - j_\beta^+ j_\beta, r_\alpha^+ s_\alpha - r_\beta^+ s_\beta]\hbar/2$$

$$= \frac{\hbar}{2\sqrt{2}} \sum_j \{\delta_{jr} j_\alpha^+ s_\alpha + \delta_{jr} j_\beta^+ s_\beta - \delta_{js} r_\alpha^+ j_\alpha - \delta_{js} r_\beta^+ j_\beta\} = 0$$

$$[S_+, q_{rs}^+(1,0)] = \sum \frac{\hbar}{\sqrt{2}} [j_\alpha^+ j_\beta, r_\alpha^+ s_\alpha - r_\beta^+ s_\beta]$$

$$= \frac{\hbar}{\sqrt{2}} \sum_j \{-\delta_{jr} j_\alpha^+ s_\beta - \delta_{js} r_\alpha^+ j_\beta\}$$

$$= \frac{\hbar 2}{\sqrt{2}} r_\alpha^+ s_\beta = \sqrt{2}\hbar q_{rs}^+(1,1)$$

$$[S_-, q_{rs}^+(1,0)] = \frac{\hbar}{\sqrt{2}} \sum_j [j_\beta^+ j_\alpha, r_\alpha^+ s_\alpha - r_\beta^+ s_\beta]$$

$$= \frac{\hbar}{\sqrt{2}} \sum_j \{\delta_{jr} j_\beta^+ s_\alpha + \delta_{js} r_\beta^+ j_\alpha\}$$

$$= \frac{2\hbar}{\sqrt{2}} r_\beta^+ s_\alpha = \sqrt{2}\hbar q_{rs}^+(1,-1)$$

We have thus shown that $q_{rs}^+(1,0)$ has the properties of a tensor operator of $S = 1$ with $M_s = 0$.

1.4 Using Eq. (1.15), the second-quantized Fermi contact Hamiltonian may be written as

$$H_F = \sum_a \sum_{rs} \frac{8\pi}{3} g\beta\gamma_a \langle \phi_r | \mathbf{S} \cdot \mathbf{I}^a \delta(\mathbf{r} - R_a) | \phi_s \rangle r^+ s$$

$$= \sum_a \sum_{rs} \frac{8\pi}{3} g\beta\gamma_a \left\langle \phi_r \left| \left[\frac{1}{2}(S_+ + S_-) I_x^a + \frac{1}{2i}(S_+ - S_-) I_y^a + S_z I_z^a \right] \delta(\mathbf{r} - R_a) \right| \phi_s \right\rangle r^+ s$$

$$= \sum_a \sum_{rs} \frac{8\pi}{3} g\beta\gamma_a \phi_r^*(R_a)\phi_s(R_a) \left(\frac{1}{2} r_\alpha^+ s_\beta I_x^a + \frac{1}{2} r_\beta^+ s_\alpha I_x^a + \frac{1}{2i} r_\alpha^+ s_\beta I_y^a - \frac{1}{2i} r_\beta^+ s_\alpha I_y^a + \frac{1}{2} r_\alpha^+ s_\alpha I_z^a - \frac{1}{2} I_z^a r_\beta^+ s_\beta \right)$$

From the definitions of $\{q^+(1,i), i = 1, 0, -1\}$ in Eq. (1.35), the final results given in the problem follow directly.

1.5

1.

$$\exp(-\boldsymbol{\kappa}) = \begin{pmatrix} 1 & 0 \\ 0 & 1 \end{pmatrix} + \begin{pmatrix} 0 & \kappa \\ -\kappa & 0 \end{pmatrix} + \frac{1}{2!}\begin{pmatrix} 0 & \kappa \\ -\kappa & 0 \end{pmatrix}\begin{pmatrix} 0 & \kappa \\ -\kappa & 0 \end{pmatrix} + \cdots$$

Since

$$\begin{pmatrix} 0 & \kappa \\ -\kappa & 0 \end{pmatrix}\begin{pmatrix} 0 & \kappa \\ -\kappa & 0 \end{pmatrix} = -\kappa^2\begin{pmatrix} 1 & 0 \\ 0 & 1 \end{pmatrix}$$

then

$$\begin{aligned}\exp(-\boldsymbol{\kappa}) &= \begin{pmatrix} 1 & 0 \\ 0 & 1 \end{pmatrix} + \kappa\begin{pmatrix} 0 & 1 \\ -1 & 0 \end{pmatrix} - \frac{\kappa^2}{2!}\begin{pmatrix} 1 & 0 \\ 0 & 1 \end{pmatrix} - \frac{\kappa^3}{3!}\begin{pmatrix} 0 & 1 \\ -1 & 0 \end{pmatrix} + \cdots. \\ &= \cos\kappa\begin{pmatrix} 1 & 0 \\ 0 & 1 \end{pmatrix} + \sin\kappa\begin{pmatrix} 0 & 1 \\ -1 & 0 \end{pmatrix} = \begin{pmatrix} \cos\kappa & \sin\kappa \\ -\sin\kappa & \cos\kappa \end{pmatrix}\end{aligned}$$

2.

$$(\tilde{\phi}_1\tilde{\phi}_2) = (\phi_1\phi_2)\begin{pmatrix} \cos\kappa & \sin\kappa \\ -\sin\kappa & \cos\kappa \end{pmatrix} = (\chi_A\chi_B)\begin{pmatrix} a_1 & a_2 \\ b_1 & b_2 \end{pmatrix}\begin{pmatrix} \cos\kappa & \sin\kappa \\ -\sin\kappa & \cos\kappa \end{pmatrix}$$

$$\kappa = 10^\circ = 0.174 \text{ rad}, \qquad \cos\kappa = 0.985, \qquad \sin\kappa = 0.174$$

$$(\tilde{\phi}_1\tilde{\phi}_2) = (\chi_a\chi_b)\begin{pmatrix} 0.985a_1 - 0.174a_2 & 0.174a_1 + 0.985a_2 \\ 0.985b_1 - 0.174b_2 & 0.174b_1 + 0.985b_2 \end{pmatrix}$$

References

Condon, E. U., and Shortley, G. H. (1935). "The Theory of Atomic Spectra." Cambridge Univ. Press, Cambridge and New York.

Dalgaard, E., and Jørgensen, P. (1978). *J. Chem. Phys.* **69**, 3833.

Davidson, E. R. (1976). "Reduced Density Matrices in Quantum Chemistry." Academic Press, New York.

Linderberg, J., and Öhrn, Y. (1973). "Propagators in Quantum Chemistry." Academic Press, New York.

Raimes, S. (1972). "Many Electron Theory." North-Holland Publ., Amsterdam.

Tinkham, M. (1964). "Group Theory and Quantum Mechanics." McGraw-Hill, New York.

Chapter 2 | Energy and Wavefunction Optimization Methods

A. INTRODUCTION

The total electronic energy of a system described by a state $|0\rangle$ is given as

$$E = \langle 0|H|0\rangle, \qquad \langle 0|0\rangle = 1 \tag{2.1}$$

In approximations commonly used to describe the true state function, $|0\rangle$ may depend on variational parameters $C_1, C_2, \ldots, C_j$, which may be expansion coefficients describing either the linear combination of configurations in $|0\rangle$ or the orbitals [Eq. (1.18)] appearing in these configurations. The total energy forms an energy hypersurface in these parameters $E(C_1, C_2, \ldots, C_j)$. We wish to determine stationary points or extrema of the energy hypersurface that, of course, occur when

$$\partial E(C_1, C_2, \ldots, C_j)/\partial C_i = 0, \qquad i = 1, 2, \ldots, j \tag{2.2}$$

In this chapter, the problem of making $E(C_1, C_2, \ldots, C_j)$ stationary will be treated for both linear and nonlinear parameters that arise in treating the most common quantum-chemical energy expressions. The first derivatives of the total energy determine the slope at a given point of the energy hypersurface, while the second derivatives of the total energy

$$\partial^2 E(C_1, C_2, \ldots, C_j)/\partial C_i \partial C_j \tag{2.3}$$

determine the curvature of the energy hypersurface and thus may be used to characterize the stationary point as a local minimum, a saddle point, or a local maximum. In attempting to find excited states of a given symmetry, one must use care to guarantee that the procedure does not permit a collapse to the lowest state of that symmetry. Procedures such as constraining the class of wavefunctions given by $\{C_i\}$ to be orthogonal to the ground state or monitoring the dominant components (largest C_i) of each wavefunction are commonly used to avoid this difficulty.

In the first optimization procedures we examine below, the parameters C_j characterize a unitary transformation of the wavefunction within the space of both orbital and configuration variations. To determine a stationary point (SP) on the energy hypersurface in this case, we derive an iterative scheme that is quadratically convergent both for ground and excited states. We use knowledge of the first and second derivatives of the total energy to determine the iterative step lengths that we have to take to reach the SP. If the energy hypersurface were parabolic in all of the parameters considered, we would reach the SP in one step. The iterative nature of the solution originates from the nonparabolic terms in the true energy hypersurface, whose description we truncate after quadratic terms.

To be more explicit about the kinds of variational parameters that commonly arise, we write the wavefunction $|0\rangle$ as a linear combination of the orthonormal basis states $\{|\phi_g\rangle\}$ that may originate from several electronic configurations:

$$|0\rangle = \sum_g |\phi_g\rangle C_{g0} \tag{2.4}$$

Each of the states $|\phi_g\rangle$ is formed from a single electronic configuration and is defined as

$$|\phi_g\rangle = \prod_{r \in g} r^+ |\text{vac}\rangle \tag{2.5}$$

where the product $\prod_{r \in g} r^+$ refers to an ordered set of creation operators. The coefficients C_{g0} are the expansion coefficients for the considered state $|0\rangle$ within this configuration basis $\{|\phi_g\rangle\}$. Variations of the spin-orbitals $\{\phi_r\}$ are commonly expressed in terms of variations in the linear expansion coefficients describing the $\{\phi_r\}$ within an atomic orbital basis. [Eq. (1.18)].

In a multiconfigurational self-consistent field (MCSCF) calculation (Dalgaard and Jørgensen, 1978; Schaefer and Miller, 1977, Chapters 3 and 4), we consider both the configuration expansion coefficients and the orbitals as variational parameters. The optimization techniques required to determine an MCSCF wavefunction are discussed in Section B. In a configuration interaction (CI) calculation, the coefficients C_{g0} are determined from Eq. (2.2) under the assumption that the orbitals are fixed. We discuss various approaches to the CI problem in more detail in Section D. The Hartree–Fock (HF) approximation assumes that the reference state refers to a single configuration but the orbitals (or creation operators) are allowed to vary and are determined from Eq. (2.2). Several techniques that have been put forth to generate optimal HF orbitals are considered in more detail in Section C.

B. MULTICONFIGURATIONAL SELF-CONSISTENT FIELD

1. Unitary Transformation of the Wavefunction

Let us now describe how one determines SPs on the energy hypersurface when the wavefunction has the form given in Eq. (2.4). We allow variations to occur in both the orbitals and the configuration expansion coefficients. In Eqs. (1.52) and (1.57) we have described how the orbital variations may be carried out by performing a unitary transformation among the orbitals. The variations in the expansion coefficients may be described in a similar manner (Dalgaard, 1980). The expansion coefficients for the state $|0\rangle$ form one column of a unitary matrix in which the remaining columns are the expansion coefficients for the orthogonal complement states within the configuration space being considered:

$$|n\rangle = \sum_{g} |\phi_g\rangle C_{gn} \tag{2.6}$$

The states $\{|0\rangle, |n\rangle\}$ and $\{|\phi_g\rangle\}$ thus are related through a unitary transformation matrix **C**. Variations in the expansion coefficients C_{gn} may be achieved either by a direct variation of these linear parameters or alternatively in terms of parameters S_{lm} describing a unitary transformation among the states $\{|l\rangle\}$. The operator

$$S = \sum_{l,m} S_{lm} |l\rangle\langle m| \tag{2.7}$$

when applied on the set of states $\{|k\rangle\}$ results in a general transformation among the states $\{|k\rangle\}$. The operator $\exp(iS)$ therefore may be used to describe a general unitary transformation among the states $\{|k\rangle\}$.

This unitary transformation shows great resemblance to the unitary transformation $\exp(i\lambda)$ in Eq. (1.36). The operator S is hermitian and the parameters S_{lm} form a hermitian matrix that determines the unitary transformation to be performed. Since we consider only real orbitals here, it becomes sufficient to use only the imaginary part of the variational parameters S_{lm}, denoted iP_{lm} [analogous to using only the $i\kappa_{rs}$ part of λ in Eq. (1.54)], and the S operator then takes the form

$$S = i \sum_{l>m} P_{lm}(|l\rangle\langle m| - |m\rangle\langle l|) \tag{2.8}$$

Further, because our interest is in optimizing the total energy for the state $|0\rangle$, we need only include the $m = 0$ parameter P_{l0} in Eq. (2.8), which then limits the operator S to be of the form

$$S = i \sum_{n \neq 0} P_{n0}(|n\rangle\langle 0| - |0\rangle\langle n|) \tag{2.9}$$

where the elements P_{n0} are real. The matrix P is a real antisymmetric matrix that, in its lower triangle, has all zero elements except for the elements P_{n0}:

$$\mathsf{P} = \begin{pmatrix} 0 & -P_{10} & -P_{20} & \cdots & -P_{n0} \\ P_{10} & 0 & 0 & \cdots & \\ P_{20} & 0 & 0 & & \\ \vdots & \vdots & & & \\ & & & & 0 \end{pmatrix} \tag{2.10}$$

The nonlinear variational parameters P_{n0} are one less in number than the linear expansion coefficients C_{g0}. This is due to the fact that a normalization condition has to be imposed on the linear expansion coefficients $\{C_{g0}\}$ if they are used as variational parameters, whereas variations described by the parameters P_{n0} automatically preserve the orthonormality of the states.

Let us now carry out the above unitary transformation. We obtain by expanding the exponential

$$\exp(iS)|m\rangle = \left[1 + iS + \frac{1}{2!}(iS)^2 + \frac{1}{3!}(iS)^3 + \cdots\right]|m\rangle. \tag{2.11}$$

The second term in the expansion may be written as

$$iS|m\rangle = ii \sum_{n\neq 0} P_{n0}(|n\rangle\langle 0| - |0\rangle\langle n|)|m\rangle = -\sum_{l} |l\rangle P_{lm}. \tag{2.12}$$

The last identity follows by the definition of the (sparse) P matrix in Eq. (2.10). The third term in the expansion in Eq. (2.11) may be determined through successive applications of Eq. (2.12) to be

$$+\tfrac{1}{2}iSiS|m\rangle = -\tfrac{1}{2}iS\sum_{l}|l\rangle P_{lm} = \tfrac{1}{2}\sum_{p,l}|p\rangle P_{pl}P_{lm} \tag{2.13}$$

Successive terms in the expansion of the exponential in Eq. (2.11) are determined in a similar manner, after which it becomes obvious that the terms may be summed to give an exponential matrix

$$\exp(iS)|m\rangle = \sum_{l}|l\rangle[\exp(-\mathsf{P})]_{lm}. \tag{2.14}$$

The actual evaluation of the exponential matrix in terms of the unitary transformation that diagonalizes $i\mathsf{P}$ may be carried out in a manner analogous to that described in Eq. (1.57) for $\exp(i\boldsymbol{\lambda})$.

Because of the especially simple nature of the above P matrix, the unitary transformation in Eq. (2.14) may be carried out analytically. We obtain by collecting together the terms arising in the $(1/n!)(iS)^n|m\rangle$ factors as sine and cosine components:

$$\exp(iS)|0\rangle = \cos x|0\rangle - \frac{1}{x}\sin x \sum_{n} P_{n0}|n\rangle \tag{2.15}$$

$$\exp(iS)|m\rangle = |m\rangle + P_{m0}\frac{1}{x}\sin x|0\rangle + \frac{1}{x^2}(\cos x - 1)P_{m0}\sum_n |n\rangle P_{n0} \quad (2.16)$$

where

$$x^2 \equiv \sum_n P_{n0}^2 \quad (2.17)$$

A unitary transformation of the reference state may now be described as

$$|\tilde{0}\rangle = \exp(i\lambda)\exp(iS)|0\rangle \quad (2.18)$$

Using the technique of Eq. (1.40) to transform all of the creation operators appearing in $|0\rangle$ and in $\exp(iS)$ (i.e., those in $|l\rangle$), we can write

$$|\tilde{0}\rangle = \exp(i\bar{S})|\bar{0}\rangle \quad (2.19)$$

where $\bar{S}$ and $|\bar{0}\rangle$ are defined as in Eqs. (2.9) and (2.4), respectively, with creation operators $\tilde{r}^+$ referring to the transformed set of orbitals. The unitary transformation of the state $|\tilde{0}\rangle$ can thus be thought of as first carrying out a unitary transformation among the orbitals in $|0\rangle$ and S and then performing a unitary transformation in the configuration space [Eq. (2.19)]. This same transformation can be viewed in a somewhat different manner. One may interpret it as first performing the configuration transformation involving all untransformed orbitals (or creation operators)

$$\exp(iS)|0\rangle = \sum_l [\exp(-\mathrm{P})]_{l0}|l\rangle \quad (2.20)$$

as given by Eq. (2.14) and then transforming the orbitals in the functions $|n\rangle$ to give

$$\exp(i\lambda)[\exp(iS)|0\rangle] = \sum_l [\exp(-P)]_{l0}|\bar{l}\rangle \quad (2.21)$$

where

$$|\bar{l}\rangle \equiv \exp(i\lambda)|l\rangle \quad (2.22)$$

Of course, both of these interpretations of Eq. (2.18) amount to nothing more than two ways of working at the same configuration and orbital transformation.

An alternative description of a unitary transformation of the reference state involves using the exponentials in Eq. (2.18) in the opposite order. This form implies that the reference state may be rewritten as

$$|\tilde{0}\rangle = \exp(iS)|\bar{0}\rangle \quad (2.23)$$

where the creation operators in $|\bar{0}\rangle$ refer to the set of transformed orbitals, while the creation operators in S correspond to the nontransformed set.

The subsequent evaluation of $\exp(iS)|\bar{0}\rangle$ would be very difficult because it would involve the computation of overlaps between states $\langle n|\bar{0}\rangle$ involving both transformed and nontransformed orbitals. This would complicate tremendously the determination of the transformed state $|\tilde{0}\rangle$; we therefore consider in the following only the unitary transformation of the reference state given in Eq. (2.18).

2. Variation of the Total Energy

The total energy corresponding to the transformed reference state is given as

$$\begin{aligned} E(\lambda, S) &= \langle 0|\exp(-iS)\exp(-i\lambda)H\exp(i\lambda)\exp(iS)|0\rangle \\ &= \langle 0|H|0\rangle - i\langle 0|[S+\lambda, H]|0\rangle + \tfrac{1}{2}\langle 0|[S,[H,S]]|0\rangle \\ &\quad + \tfrac{1}{2}\langle 0|[\lambda,[H,\lambda]]|0\rangle + \langle 0|[S,[H,\lambda]]|0\rangle + \cdots \end{aligned} \tag{2.24}$$

By introducing a matrix notation in which the variational parameters κ_{rs} and P_{n0} form row and column vectors, we can rewrite Eq. (2.24) as

$$E(\lambda, S) = E(0,0) - 2(\boldsymbol{\kappa}\mathbf{P})\begin{pmatrix}\mathbf{W}\\ \mathbf{V}\end{pmatrix} + (\boldsymbol{\kappa}\mathbf{P})(\mathsf{A}-\mathsf{B})\begin{pmatrix}\boldsymbol{\kappa}\\ \mathbf{P}\end{pmatrix} + \cdots \tag{2.25}$$

We have introduced in Eq. (2.25) the short-hand notation for the operators

$$Q^{+} = \{r^{+}s\}(r > s), \qquad R^{+} = \{|n\rangle\langle 0|\} \tag{2.26}$$

and defined the matrices

$$\mathbf{W} = \langle 0|[Q,H]|0\rangle \tag{2.27}$$

$$\mathbf{V} = \langle 0|[R,H]|0\rangle \tag{2.28}$$

$$\mathsf{A} = \begin{pmatrix}\mathsf{A}_{11} & \mathsf{A}_{12}\\ \mathsf{A}_{21} & \mathsf{A}_{22}\end{pmatrix} = \begin{pmatrix}\langle 0|[Q,H,Q^{+}]|0\rangle & \langle 0|[[Q,H],R^{+}]|0\rangle\\ \langle 0|[R,[H,Q^{+}]]|0\rangle & \langle 0|[R,H,R^{+}]|0\rangle\end{pmatrix} \tag{2.29}$$

$$\mathsf{B} = \begin{pmatrix}\mathsf{B}_{11} & \mathsf{B}_{12}\\ \mathsf{B}_{21} & \mathsf{B}_{22}\end{pmatrix} = \begin{pmatrix}\langle 0|[Q,H,Q]|0\rangle & \langle 0|[[Q,H],R]|0\rangle\\ \langle 0|[R,[H,Q]]|0\rangle & \langle 0|[R,H,R]|0\rangle\end{pmatrix} \tag{2.30}$$

For convenience, we have introduced the double commutator, defined as

$$[Q,H,Q^{+}] = \tfrac{1}{2}\{[Q,[H,Q^{+}]] + [[Q,H],Q^{+}]\} \tag{2.31}$$

which arises naturally in A_{11}, A_{22}, B_{11}, and B_{22} because

$$\langle 0|[\lambda,[H,\lambda]]|0\rangle - \langle 0|[[\lambda,H],\lambda]|0\rangle = \langle 0|[[\lambda,\lambda],H]|0\rangle = 0 \tag{2.32}$$

and an analogous result for S. The matrices **W**, **V** determine the first-order variations of the energy function, which at a SP on the energy hypersurface are zero. The condition that $\mathbf{V}$ and $\mathbf{W}$ are zero at a SP is often referred to as the generalized Brillouin theorem (GBT).

The matrix $\mathsf{A} - \mathsf{B}$ defines the second-order variation of the energy function and is often referred to as the Hessian matrix. The double-commutator form of the Hessian matrix allows these second-order terms to be expressed as a quadratic form.

3. One-Step Second-Order Procedure

As stated previously, a SP on the energy hypersurface is obtained when $\delta E(\lambda, S) = 0$. Neglecting third- and higher-order terms in the energy function [which rigorously no longer makes $E(\lambda, S)$ a true expectation value] we obtain from Eq. (2.25), by differentiating with respect to $\boldsymbol{\kappa}$ and $\mathbf{P}$,

$$-\begin{pmatrix} \mathbf{W} \\ \mathbf{V} \end{pmatrix} + (\mathsf{A} - \mathsf{B})\begin{pmatrix} \boldsymbol{\kappa} \\ \mathbf{P} \end{pmatrix} = \mathbf{0} \tag{2.33}$$

or equivalently

$$\begin{pmatrix} \boldsymbol{\kappa} \\ \mathbf{P} \end{pmatrix} = (\mathsf{A} - \mathsf{B})^{-1}\begin{pmatrix} \mathbf{W} \\ \mathbf{V} \end{pmatrix} \tag{2.34}$$

as the conditions for a SP. The matrices $\boldsymbol{\kappa}$ and $\mathbf{P}$ may then be determined from Eq. (2.34) and a set of transformed orbitals and states obtained from Eqs. (1.52) and (2.14), respectively. If the energy hypersurface contained no higher than quadratic terms, we would reach a SP in one iteration of the above procedure. The third- and higher-order terms in the energy function do, however, require that an iterative scheme be applied to determine a SP. The iterative scheme may be described as follows: From an initial guess of orbitals and a choice of the configuration space, we determine a set of approximate eigenstates $|n\rangle$ (e.g., by performing a configuration interaction calculation). The matrices $\mathbf{V}$, $\mathbf{W}$, A, and B are then determined and Eq. (2.34) is solved to give the matrices $\boldsymbol{\kappa}$ and $\mathbf{P}$. A transformed set of orbitals and states may then be obtained from Eqs. (1.52) and (2.14) and the procedure repeated until the numerical values of $\mathbf{W}$ and $\mathbf{V}$ are smaller than a specific tolerance. The above described approach has included all terms in the energy function through second order and is therefore quadratically convergent. We therefore denote this scheme the one-step second-order approach (Yeager and Jørgensen, 1979).

4. Two-Step Procedure

Another approach, which differs slightly in its realization of the iterative procedure, has also been used and is referred to here as the two-step second-order scheme. It may be described as follows: After an initial guess of orbitals, a configuration interaction calculation (see Section D) is carried out to determine the starting set of CI eigenstates $|l\rangle$. We then have

$$\langle m|H|l\rangle = E_l \delta_{ml} \tag{2.35}$$

and the **V** matrix

$$V_n = \langle 0|[R_n, H]|0\rangle = \langle 0|[|0\rangle\langle n|, H]|0\rangle = 0 \tag{2.36}$$

becomes equal to zero. Equation (2.34) may then be partitioned (Löwdin, 1968) to give (using $\mathsf{B}_{22} = 0$)

$$\boldsymbol{\kappa} = [\mathsf{A}_{11} - \mathsf{B}_{11} - (\mathsf{A}_{12} - \mathsf{B}_{12})\mathsf{A}_{22}^{-1}(\mathsf{A}_{21} - \mathsf{B}_{21})]^{-1}\mathbf{W} \tag{2.37}$$

and the $\boldsymbol{\kappa}$ matrix can be determined from this set of linear equations.

A transformed set of orbitals may now be obtained using this $\boldsymbol{\kappa}$ in Eq. (1.52) and a new CI calculation (diagonalization of $\langle l|H|m\rangle$) carried out. This process is then continued until convergence is reached. In the two-step second-order procedure, Eq. (2.34) is thus always applied in a basis where the states are determined from a CI calculation. The matrix **P** is never explicitly calculated. In contrast, in the one-step procedure the configuration expansion coefficients of $|0\rangle$ and $|n\rangle$ are determined from the unitary transformation given in Eq. (2.14), where **P** is obtained from Eq. (2.34) rather than from a CI calculation.

The terms $\mathsf{A}_{12} - \mathsf{B}_{12}$ coupling the configuration and orbital space variation have been neglected in many calculations. In many cases, these terms show little effect on the convergence rate of the procedure. It should, however, be pointed out that a quadratically convergent scheme is only obtained when these coupling terms are included.

5. Explicit Hessian and Generalized Brillouin Matrix Elements

Let us now consider the evaluation of some of the matrix elements appearing above. In the one-step procedure we have to calculate

$$(A_{22})_{mn} = \langle 0|[R_m, H, R_n^+]|0\rangle = \langle m|H|n\rangle - \delta_{mn}\langle 0|H|0\rangle \tag{2.38}$$

and

$$V_n = \langle 0|[R_n, H]|0\rangle = \langle n|H|0\rangle \tag{2.39}$$

The elements A_{22} and **V** thus contain all matrix elements contained in a configuration interaction calculation within the considered configuration space. When the iterative MCSCF procedure has converged, all elements of **V** are zero and the interactions between the reference state $|0\rangle$ and the residual states are thus eliminated. The diagonal and off-diagonal matrix elements of the Hamiltonian in the residual space $\{|n\rangle\}$ may, however, all be non-vanishing.

In the two-step second-order approach, the CI calculation that is carried out in each step prior to the evaluation of the matrices simplifies the evaluation of $\mathbf{V}$ and A_{22}. The $\mathbf{V}$ matrix becomes, as stated earlier, zero and the A_{22} matrix

$$(A_{22})_{mn} = \delta_{mn}(E_m - E_0) \tag{2.40}$$

becomes diagonal. These simplifications remain in each step of the iterative process because a CI calculation is performed in each iteration.

Except for A_{22} and $\mathbf{V}$, the form of the matrix element in the one- and two-step procedures are the same. The matrix elements of $\mathbf{W}$, A_{11}, and B_{11} may be derived from Eqs. (2.41) and (2.42) by index substitution. The excitation operators in these equations have singlet spin symmetry, since they arise in the operator λ, which must preserve the symmetry of $|0\rangle$ in forming $\exp(i\lambda)|0\rangle$. These matrices can be expressed in terms of one- and two-electron integrals and the one- and two-electron density matrices as given below. Note that no more than two-electron density matrices appear in $\mathbf{W}$, A_{11}, and B_{11}:

$$\begin{aligned}\langle 0|[t_\alpha^+ u_\alpha + t_\beta^+ u_\beta, H]|0\rangle &= \sum_{\sigma p} h_{up}\langle 0|t_\sigma^+ p_\sigma|0\rangle - \sum_{\sigma p} h_{pt}\langle 0|p_\sigma^+ u_\sigma|0\rangle \\ &\quad - \sum_{pqr}\langle pq|rt\rangle\rho_{qpru} + \sum_{qrs}\langle uq|rs\rangle\rho_{tqsr}\end{aligned} \tag{2.41}$$

$$\begin{aligned}&\langle 0|[l_\alpha^+ k_\alpha + l_\beta^+ k_\beta, [H, t_\alpha^+ u_\alpha + t_\beta^+ u_\beta]]|0\rangle \\ &\quad = h_{kt}\sum_\sigma \langle 0|l_\sigma^+ u_\sigma|0\rangle + h_{ul}\sum_\sigma \langle 0|t_\sigma^+ k_\sigma|0\rangle - \delta_{kt}\sum_{p\sigma} h_{up}\langle 0|l_\sigma^+ p_\sigma|0\rangle \\ &\qquad - \delta_{lu}\sum_{p\sigma} h_{pt}\langle 0|p_\sigma^+ k_\sigma|0\rangle - \delta_{lu}\sum_{pqr}\langle pq|rt\rangle\rho_{pqkr} - \delta_{kt}\sum_{qrs}\langle uq|rs\rangle\rho_{lqsr} \\ &\qquad - \sum_{pq}\langle pq|lt\rangle\rho_{pquk} - \sum_{rs}\langle uk|rs\rangle\rho_{tlsr} + \sum_{pr}\langle kp|rt\rangle\rho_{lpur} \\ &\qquad + \sum_{pr}\langle kp|tr\rangle\rho_{plur} + \sum_{qs}\langle uq|ls\rangle\rho_{tqsk} + \sum_{qs}\langle uq|sl\rangle\rho_{qtsk}\end{aligned} \tag{2.42}$$

where

$$\rho_{ijkl} = \sum_{\sigma,\sigma'}\langle 0|i_\sigma^+ j_{\sigma'}^+ k_{\sigma'} l_\sigma|0\rangle \tag{2.43}$$

and σ and σ' run over the electron spin indices α and β.

The elements of A_{21} and B_{21} reduce as follows:

$$\langle 0|[|0\rangle\langle n|, [H, t_\alpha^+ u_\alpha + t_\beta^+ u_\beta]]|0\rangle = \langle n|[H, t_\alpha^+ u_\alpha + t_\beta^+ u_\beta]|0\rangle \tag{2.44}$$

and an explicit final formula for Eq. (2.44) may be obtained from Eq. (2.41) by replacing the one- and two-electron density matrices with the corresponding transition density matrix elements.

6. Mode Damping

The GBT matrix and the Hessian matrix arising in the one-step second-order procedure determine the energy slope and curvature, respectively, for a given point on the hypersurface. When a SP point has been reached, the eigenvalues of the Hessian matrix thus can be used to characterize this point. We have reached a local minimum if all eigenvalues are positive. Mixed positive and negative eigenvalues correspond to a saddle point on the energy hypersurface. In employing the two-step procedure outlined above, one no longer has the opportunity to characterize the state by its Hessian eigenvalues, because the full Hessian matrix is not employed and the partitioned Hessian of Eq. (2.37) does not have the same eigenvalues as the full Hessian.

Some insight into the step lengths $(\boldsymbol{\kappa}, \mathbf{P})$ that should be taken in second-order procedures may be obtained by transforming the second-order equation to a form in which the Hessian matrix is diagonal. Let us consider initially the diagonalization (by the unitary matrix U) of the full $\mathsf{A} - \mathsf{B}$ matrix appearing in the one-step second-order equation

$$\mathsf{A} - \mathsf{B} = \mathsf{U}\boldsymbol{\varepsilon}\mathsf{U}^{+} \tag{2.45}$$

Equation (2.34) then becomes

$$\begin{pmatrix} \tilde{\boldsymbol{\kappa}} \\ \tilde{\mathbf{P}} \end{pmatrix} = \boldsymbol{\varepsilon}^{-1} \begin{pmatrix} \tilde{\mathbf{W}} \\ \tilde{\mathbf{V}} \end{pmatrix} \tag{2.46}$$

where

$$\begin{pmatrix} \tilde{\boldsymbol{\kappa}} \\ \tilde{\mathbf{P}} \end{pmatrix} = \mathsf{U}^{+} \begin{pmatrix} \boldsymbol{\kappa} \\ \mathbf{P} \end{pmatrix} \tag{2.47}$$

$$\begin{pmatrix} \tilde{\mathbf{W}} \\ \tilde{\mathbf{V}} \end{pmatrix} = \mathsf{U}^{+} \begin{pmatrix} \mathbf{W} \\ \mathbf{V} \end{pmatrix} \tag{2.48}$$

Each normal mode on the energy hypersurface is decoupled and hence may be described independently. This is particularly useful in the initial iterations of an MCSCF calculation, where third- and higher-order terms may be important and even dominate as a result of the poor initial guess of the orbitals. The second-order scheme may, in such cases, be forced to take step lengths $(\tilde{\boldsymbol{\kappa}}, \tilde{\mathbf{P}})$ that are too large. The normal mode analysis of Eq. (2.46), which displays the slopes $(\tilde{\mathbf{W}}, \tilde{\mathbf{V}})$ and curvature $(\boldsymbol{\varepsilon})$ of each mode independently, then becomes a convenient tool to use for changing the step length for those modes that take very large steps. This is done by restricting the allowed size of the $\tilde{\kappa}$ and $\tilde{\mathbf{P}}$ matrix elements. In ground-state calculations where the Hessian matrix has to be positive definite, we may even change the direction of the step (i.e., change the sign of $\tilde{\kappa}$) if small negative eigenvalues ε_j appear. This situation occurs frequently in the initial iterations of actual calculations. If the matrix elements that couple the orbital and coefficient optimization (the $\mathsf{A}_{12} - \mathsf{B}_{12}$ matrix) are very small, $\tilde{\kappa}$ then predominantly refers to the orbital optimization while $\tilde{\mathbf{P}}$ refers to the coefficient optimization. In these cases, it is reasonable to impose some different limits upon the size of the maximum elements of the step length vectors $\tilde{\kappa}$ and $\tilde{\mathbf{P}}$. At present, there is little experience on how to optimally make these restrictions although results of initial calculations indicate that the basic philosophy is correct. When strong coupling occurs between the configuration and the orbital space, more refined damping schemes may need to be introduced (Yeager *et al.*, 1980).

In the two-step second-order procedure, damping may only be performed in the space that is dominated by the orbital space. From applying the unitary transformation to Eq. (2.37) we get

$$\tilde{\kappa} = \varepsilon^{-1}\mathbf{W} \tag{2.49}$$

where

$$\mathsf{A}_{11} - \mathsf{B}_{11} - (\mathsf{A}_{12} - \mathsf{B}_{12})A_{22}^{-1}(\mathsf{A}_{21} - \mathsf{B}_{21}) = \mathsf{U}\varepsilon\mathsf{U}^{+} \tag{2.50}$$

$$\tilde{\kappa} = \mathsf{U}^{+}\kappa \tag{2.51}$$

$$\tilde{\mathbf{W}} = \mathsf{U}^{+}\mathbf{W} \tag{2.52}$$

Because the reference state $|0\rangle$ and its orthogonal complement states $|n\rangle$ are determined from a CI calculation, it is not generally possible to impose constraints on the step lengths in the configuration space. Further, the CI steps are not necessarily taken along the normal modes. In particular, when strong coupling elements exist between the configuration and orbital spaces, large fluctuations in the amplitude of the dominant configuration may be encountered, which may lead to difficulties in converging to the state under consideration.

7. Elimination of Redundant Operators

Having now given a general discussion of quadratically convergent second-order MCSCF methods together with some analysis of how such techniques might best be implemented, we can move on to describe other MCSCF methods, as well as to give more detail about the numerical requirements of such calculations. Before doing so, however, it is important that we address a technical point that must be understood if one is to be successful in carrying out such MCSCF calculations.

The unitary transformation of the reference state given in Eq. (2.18) has as generators the operators r^+s of λ and $|n\rangle\langle 0|$ of S. It is possible that the operators r^+s and $|n\rangle\langle 0|$ span the same space. That is, the effects of the operators r^+s may be expressed in terms of those of the state projections in the configuration space. To determine whether the effects of a given operator r^+s can be expressed in terms of the kets $\{|m\rangle\}$, we examine the following difference ket:

$$|f\rangle \equiv r^+s|l\rangle - \sum_m |m\rangle\langle m|r^+s|l\rangle \tag{2.53}$$

If the norm of $|f\rangle$ vanishes, then $|f\rangle$ itself vanishes and hence $r^+s|l\rangle$ can be exactly represented as a sum of the $\{|m\rangle\}$ functions. The norm of $|f\rangle$ vanishes when

$$\langle f|f\rangle = 0 = \langle l|s^+rr^+s|l\rangle - \sum_m \langle l|s^+r|m\rangle\langle m|r^+s|l\rangle \tag{2.54}$$

or, in other words, when

$$\sum_m |\langle m|r^+s|l\rangle|^2 = \langle l|s^+rr^+s|l\rangle \tag{2.55}$$

When *both* the operators r^+s and s^+r fulfill Eq. (2.55), for any state $|l\rangle$ the variations described by the parameters λ_{rs} will be denoted as redundant.

The search for redundant variables may, of course, alternatively be performed in the configuration space $\{|\phi_g\rangle\}$ since this space is related to the space $\{|l\rangle\}$ through a unitary transformation. Because the states $\{r^+s|\phi_g\rangle\}$ are normalized to unity, the search for redundant variables may be achieved by investigating whether the sum

$$\sum_{g'} |\langle \phi_{g'}|r^+s|\phi_g\rangle|^2 \tag{2.56}$$

is equal to zero or one for any state $|\phi_g\rangle$.

We now show how orbital changes caused by redundant variables can be represented as configuration changes caused by S and can thus be eliminated from the energy optimization procedure. The redundant set of operators form a hermitian operator

$$\lambda'' = \sum_{rs} \lambda''_{rs} r^+s \tag{2.57}$$

The operator λ', which contains all of the r^+s that are not redundant also forms a hermitian operator

$$\lambda' = \sum_{rs} \lambda'_{rs} r^+s \tag{2.58}$$

Since the operators in Eq. (2.57) are generators of a subgroup of the unitary group, $\exp(i\lambda)$ may be factorized to give

$$\exp(i\lambda) = \exp(i\lambda')\exp(i\lambda'') \tag{2.59}$$

Equation (2.59) is thus a representation of an arbitrary group element expressed as a left coset of this subgroup. Expressed in other words, the unitary transformation that is described by $\exp(i\lambda)$ may alternatively be described by the unitary transformation $\exp(i\lambda')\exp(i\lambda'')$. It should be pointed out that there exist no simple relations between the λ_{rs} parameters and the λ'_{rs} and λ''_{rs} parameters. With the above factorization of the "redundant" part (λ''), the unitary transformation of the reference state may be written as

$$|\tilde{0}\rangle = \exp(i\lambda')\exp(i\lambda'')\exp(iS)|0\rangle$$

Since $|f\rangle$ in Eq. (2.53) is zero for any product of redundant operators,

$$\exp(i\lambda'')|l\rangle = \sum_p |p\rangle\langle p|\exp(i\lambda'')|l\rangle \tag{2.60}$$

Using this relation together with Eq. (2.14) gives

$$|\tilde{0}\rangle = \exp(i\lambda')\exp(i\lambda'')\sum_l |l\rangle[\exp(-P)]_{l0}$$

$$= \exp(i\lambda')\sum_{p,l} |p\rangle\langle p|\exp(i\lambda'')|l\rangle[\exp(-P)]_{l0} \tag{2.61}$$

The matrix $\{\langle p|\exp(i\lambda'')|l\rangle\}$ is unitary since the scalar product of Eq. (2.60) with $\langle q|\exp(-i\lambda'')$ gives

$$\sum_p \langle q|\exp(-i\lambda'')|p\rangle\langle p|\exp(+i\lambda'')|l\rangle = \delta_{ql}$$

Therefore, the product matrix

$$\sum_l \langle p|\exp(i\lambda'')|l\rangle \cdot [\exp(-P)]_{l0} \tag{2.62}$$

must consequently also be unitary. Because a single unitary transformation of the form given in Eqs. (2.10) and (2.14) is sufficient for optimizing the total energy, the redundant variables may be left out when optimizing the energy. That is, the λ'' factors can do nothing more, in a wavefunction optimization, than can be done by the $\exp(iS)$ operator.

8. Practical Considerations

So far, no attention has been given to the spatial and spin symmetry features of the reference state. The theory we have outlined thus far may hence be described as unrestricted multiconfigurational HF. In most applications (Eyring *et al.*, 1967), we require the reference state to have a certain symmetry (i.e., the reference state should transform according to an irreducible representation of the Hamiltonian's point group). For the wavefunction symmetry to be conserved under a sequence of unitary transformations, the operators λ and S have to be tensor operators belonging to the totally symmetric irreducible representation. In this way, the symmetry of the wavefunction would be conserved during the iteration procedure.

Calculations of the matrix elements that are used to define the above procedures requires knowledge of the one- and two-electron integrals in the MCSCF spin-orbital basis. Therefore, a two-electron integral transformation (Schaefer and Miller, 1977, Chapt. 6) has to be performed in each step of the iterative procedure. MCSCF approaches, in general, require such repeated two-electron integral transformations to be performed. Since these transformations may, in many cases, be the computationally most demanding step of the calculation, it becomes very important to use MCSCF procedures that converge reliably in a minimum number of iterations. We have chosen to emphasize here the one- and two-step second-order procedures because they are quadratically convergent and because they allow a controlled (damped) "walk" to be performed on the energy hypersurface when cubic and higher-order terms and/or coupling between orbitals and configuration optimizations are important.

9. Generalized Brillouin-Theorem-Based Procedures

So far we have used the condition that the energy function be stationary to define MCSCF schemes. The existence of a stationary point on the energy hypersurface requires that the GBT be fulfilled at this point. Hence, iterative MCSCF procedures may alternatively be developed by insisting that the GBT be satisfied as the iterative procedure converges. A quadratically convergent scheme may be obtained by further insisting that the error in the GBT matrix in the $(n+1)$th iteration should be the square of the error in the nth iteration. Denoting the operators and states in the $(n+1)$th iteration with a tilde and those of the n'th with no tilde, using Eqs. (2.18) and (1.38) we obtain

$$\mathbf{W}_{n+1} = \langle\tilde{0}|[\tilde{Q},H]|\tilde{0}\rangle = \langle 0|[Q,H]|0\rangle + i\langle 0|[Q,[H,\lambda]]0\rangle + i\langle 0|[[Q,H],S]|0\rangle + O(\kappa^2,P^2) \tag{2.63}$$

$$\mathbf{V}_{n+1} = \langle\tilde{0}|[\tilde{R},H]|\tilde{0}\rangle = \langle 0|[R,H]|0\rangle + i\langle 0|[R,[H,\lambda]]|0\rangle + i\langle 0|[R,[H,S]]|0\rangle + O(\kappa^2,P^2) \tag{2.64}$$

since, for example,

$$\begin{aligned}\langle 0|[Q,[H,\lambda]]|0\rangle - \langle 0|[[Q,H],\lambda]|0\rangle &= \langle 0|[[Q,\lambda],H]|0\rangle \\ &= \boldsymbol{\kappa}\langle 0|[[Q,Q-Q^+],H]|0\rangle \\ &= O(\kappa^2)\end{aligned} \tag{2.65}$$

The last identity arises because the GBT is not fulfilled until convergence is reached, and thus $\langle 0|[[Q, Q - Q^+], H]|0\rangle$ is of order κ itself. The double commutator may be introduced in Eq. (2.63), and Eqs. (2.63) and (2.64) may then be combined to give

$$\begin{pmatrix} \mathbf{W} \\ \mathbf{V} \end{pmatrix}_{n+1} = \begin{pmatrix} \mathbf{W} \\ \mathbf{V} \end{pmatrix}_{n} - (\mathsf{A} - \mathsf{B})_n \begin{pmatrix} \boldsymbol{\kappa} \\ \mathbf{P} \end{pmatrix} + O(\kappa^2, P^2) \tag{2.66}$$

where we have used Eqs. (2.27)–(2.30). A quadratically convergent scheme is thus obtained when the $\boldsymbol{\kappa}$ and $\mathbf{P}$ matrices are determined from

$$\begin{pmatrix} \mathbf{W} \\ \mathbf{V} \end{pmatrix}_{n} = (\mathsf{A} - \mathsf{B})_n \begin{pmatrix} \boldsymbol{\kappa} \\ \mathbf{P} \end{pmatrix} \tag{2.67}$$

which is identical to the one-step second-order equation [Eq. (2.33)]. Hence the one-step second-order procedure described earlier can also be viewed as arising from the GBT.

Most MCSCF procedures that have been employed to date (Schaefer and Miller, 1977, Chapters 3 and 4) have concentrated on deriving iterative schemes based upon only insisting that

$$\langle 0|[H, r^+s]|0\rangle = 0 \tag{2.68}$$

in each step of the iterative procedure. As successive sets of MCSCF orbitals are determined in each step of the iterative procedure, the configuration space equivalent of the GBT $\langle 0|[H, |n\rangle\langle 0|]|0\rangle = 0$ is achieved through performing a CI calculation within the limited configuration space.

To see how Eq. (2.68) can be used to define an iterative process, let us consider the first two terms in the expansion of $\exp(i\lambda)|0\rangle$:

$$|0\rangle - \sum_{r>s} \kappa_{rs}(r^+s - s^+r)|0\rangle \tag{2.69}$$

This first-order approximation to the true $\exp(i\lambda)|0\rangle$ then leads us to consider the variational wavefunction

$$|\tilde{0}\rangle \simeq X_0|0\rangle + \sum_{r>s} X_{rs}(r^+s - s^+r)|0\rangle \tag{2.70}$$

containing the linear variational parameters X_0 and $\{X_{rs}\}$. The optimal values of these parameters may then be determined from the superconfiguration interaction (SCI) secular problem (Banerjee and Grein, 1976)

$$\mathsf{H}\mathbf{X} = E\mathsf{S}\mathbf{X} \tag{2.71}$$

The SCI Hamiltonian matrix elements are defined as

$$H_{0,rs} = \langle 0|H(r^+s - s^+r)|0\rangle \tag{2.72}$$

which for real orbitals reduces to Eq. (2.68):

$$H_{0,rs} = \langle 0|[H, r^+s]|0\rangle = H_{rs,0} \tag{2.73}$$

The other matrix elements of H are

$$H_{00} = \langle 0|H|0\rangle \tag{2.74}$$

$$H_{rs,tu} = \langle 0|(r^+s - s^+r)^+ H(t^+u - u^+t)|0\rangle \tag{2.75}$$

The scalar product matrix S is defined in a similar fashion (e.g., $S_{0,rs} = \langle 0|r^+s - s^+r|0\rangle = 0$) and the eigenvector **X** has the components $\mathbf{X} = \{X_0, X_{rs}\}$. The GBT therefore states that in the so-called SCI secular problem [Eq. (2.71)], the state $|0\rangle$ should be noninteracting with its single excitations $(r^+s - s^+r)|0\rangle$. Once this occurs, Eq. (2.71) will have, as one of its eigenvalues, the MCSCF energy $\langle 0|H|0\rangle$. The other eigenvalues, as in all variational secular problems, represent upper bounds to other true energy levels.

The eigenvector **X** obtained from the SCI secular problem can be used to define a transformation of the orbital appearing in $|0\rangle$. To see how this transformation arises, we rewrite Eq. (2.70) as

$$|\tilde{0}\rangle = X_0^{1-N}\left[X_0^N + \sum_{r>s} X_0^{N-1} X_{rs}(r^+s - s^+r)\right]\sum_g C_{g0}|\phi_g\rangle \tag{2.76}$$

The effect of $\sum_{r>s}(r^+s - s^+r)X_{rs}$ on each configuration $|\phi_g\rangle$ results in two new configurations in which spin orbital ϕ_s is replaced by ϕ_r and vice versa. For example, the effect on $1^+2^+ \cdots N^+|\text{vac}\rangle$ is to give

$$\sum_{i=1}^{N}\left[\sum_{r>i} X_{ri}1^+2^+ \cdots (i-1)^+ r^+ (i+1)^+ \cdots N^+ \right.$$
$$\left. - \sum_{r<i} X_{ir}1^+2^+ \cdots (i-1)^+ r^+ (i+1)^+ \cdots N^+\right]|\text{vac}\rangle \tag{2.77}$$

If the spin-orbitals occupied in any configuration $|\phi_g\rangle$ are denoted by ϕ_t, then the above SCI wavefunction in Eq. (2.76) can be expressed as

$$|\tilde{0}\rangle = X_0^{1-N}\sum_g C_{g0}\prod_{t\in g}\left[X_0 t^+ + \sum_{r>t} X_{rt} r^+ - \sum_{r<t} X_{tr} r^+\right]|\text{vac}\rangle + O(X_{rs}^2) \tag{2.78}$$

That is, the wavefunction used in the SCI calculation (Eq. (2.70)) is identical, through first order in the X_{rs} parameters, to a new linear combination of configurations with the same C_{g0} coefficients but with orbitals $\tilde{\phi}_t$ that can be expressed in terms of the original orbitals as

$$\tilde{\phi}_t = X_0\phi_t + \sum_{r>t} X_{rt}\phi_r - \sum_{r<t} X_{tr}\phi_r \tag{2.79}$$

Because this orbital transformation is not properly unitary (the $\{\tilde{\phi}_k\}$ are normalized and orthogonal only through first order in the X_{rs}), the set $\{\tilde{\phi}_k\}$ must, in each iteration, be orthonormalized (by, for example, the Schmidt or Löwdin procedure).

The SCI iterative procedure thus consists of guessing a starting set of orbitals and generating the $\{C_{g0}\}$ expansion coefficients from a CI calculation. The SCI secular problem is then constructed and solved (to give **X**) after which the new orbitals $\{\tilde{\phi}_k\}$ are computed as in Eq. (2.79) and subsequently orthonormalized. These new orbitals are then used to perform a new CI calculation to generate new $\{C_{g0}\}$ coefficients and hence a new SCI secular problem. This iterative procedure is continued until convergence is achieved at which time the GBT is fulfilled. A significant drawback of most SCI procedures as now implemented is that they do not treat the coupling between orbital and configuration optimization. SCI methods that treat both optimizations on equal footing represent a significant improvement. In situations for which strong coupling exists between the orbital and configuration space, the above-described two-step SCI process might thus be expected to converge slowly. As we mentioned above, the quadratically convergent one-step second order procedure discussed in the preceding section could also be viewed as being defined, through Eq. (2.66), to make the GBT obeyed. It is then important to explore how the two iterative methods, both of which can be stated through the GBT, differ. The difference arises from terms in

$$\langle 0|\exp(-iS)\exp(-i\lambda)H\exp(i\lambda)\exp(iS)|0\rangle$$

that are quadratic in S or λ and that arise from the second-order components of the individual exponential operators. For example, $\langle 0|Hi\lambda i\lambda|0\rangle$ and $\langle 0|iSiSH|0\rangle$ arise in the exponential formulation but do not arise in the expectation value of the SCI wavefunction given in Eq. (2.69). The neglect of second-order terms and the requisite reorthogonalization of the MCSCF orbitals differentiate between the two methods and render the SCI approach not quadratically convergent.

Because SCI approaches to the MCSCF problem are not based upon extremizing the full second-order energy expression described above, their convergence rate is linear rather than quadratic, although in practice such SCI methods may sometimes demonstrate approximate quadratic convergence. Because the SCI energies result from solutions of an eigenvalue problem, each SCI energy is an upper bound to the respective true energies (ground and excited). The values of X_{ij} obtained from the SCI secular problem [Eq. (2.71)] when used to carry out orbital modifications [through Eq. (2.79)] yield a new multiconfigurational wavefunction whose Hamiltonian expectation value is, because of the subsequent orthonormalization needed, no longer identical to the eigenvalue E that was obtained from the SCI secular problem.

In the unitary second-order method, the energy expression $E(\lambda, S)$ given in Eq. (2.25), when truncated after terms linear and quadratic in λ and S, is no longer an expectation value of H and thus no longer bounds the ground-state–total energy. Thus, the stationary points of $E(\lambda, S)$ do not form rigorous upper bounds to the respective true ground- and excited-state energies. Of course, there are good reasons to believe that, in the neighborhood of an eigenstate, $E(\lambda, S)$ can be well approximated by this quadratic hypersurface. Moreover, the values of $\mathbf{P}$ and $\boldsymbol{\kappa}$ obtained from making $E(\lambda, S)$ stationary, when used in Eqs. (1.52) and (2.14) to obtain $|\tilde{0}\rangle$, do give a proper upper-bound energy through $\langle\tilde{0}|H|\tilde{0}\rangle$.

Having now discussed how one can go about optimizing the electronic energy of an MCSCF wavefunction, we turn our attention to two special subclasses of this procedure; the single-configuration SCF problem and the frozen-orbital CI problem. Because we choose to view these situations as special cases of the above MCSCF problem, we obtain a specialized view of SCF and CI theory. There already exist in the literature extensive and clear treatments of SCF and CI as they are more commonly treated within the linear variational framework. Hence we have not attempted to cover the more conventional aspects of these topics here.

C. SINGLE-CONFIGURATION SELF-CONSISTENT FIELD METHODS

1. Quadratically Convergent Scheme

Let us consider a situation in which we choose to work with a one-configuration wavefunction for which the orbitals are allowed to vary. This single configuration $|0\rangle$ may still consist of a linear combination of determinants whose (fixed) coefficients are determined by the space and spin symmetry imposed on $|0\rangle$. The orbital variations may be described by $\exp(i\lambda)$ and an optimal set of orbitals determined as in the previous section [by simply neglecting terms involving $\exp(iS)$]. The second-order Eq. (2.33) then reads

$$\mathbf{W} = (\mathsf{A}_{11} - \mathsf{B}_{11})\boldsymbol{\kappa} \tag{2.80}$$

where A_{11} and B_{11} are defined in Eqs. (2.29) and (2.30). A quadratically convergent scheme for optimizing orbitals may be described as follows. Given an initial guess for the "occupied" orbitals, we use Eq. (2.80) to determine $\boldsymbol{\kappa}$, and then we use Eq. (1.52) to generate a transformed set of orbitals. This process is repeated until convergence is reached. This process requires, even in the one-configuration case, a partial two-electron integral transformation in each step of the iterative procedure. For a single-configuration case, a more restricted two-electron integral transformation can be used in each step if, instead of the above quadratic procedure, one uses an approach that is based on the Brillouin condition alone. These so-called first-order BT-based self-consistent-field (SCF) procedures are, however, not quadratically convergent, much as the SCI method treated earlier is only a linearly convergent MCSCF method.

2. Brillouin-Theorem-Based Methods

The HF or SCF approaches based upon the BT itself,

$$\langle 0|[H, r^+s]|0\rangle = 0 \tag{2.81}$$

introduce a decomposition of the Hamiltonian into a Fock operator (which the spin-orbital basis is chosen to diagonalize)

$$F = \sum_{r,s}(h_{rs} + V_{rs})r^+s = \sum_r \varepsilon_r r^+ r \tag{2.82}$$

where h_{rs} is the one-electron part of the Hamiltonian. A Fock potential

$$V = \sum_{r,s} V_{rs} r^+ s \tag{2.83}$$

and the electron repulsion term W combine with F so that

$$H = F - V + W \tag{2.84}$$

The one-electron Fock potential V is thus far arbitrary. Different choices for V correspond to different choices of the spin-orbitals $\{\phi_r\}$ and their corresponding orbital energies $\{\varepsilon_r\}$, since we require the ϕ_r and ε_r to obey

$$h_{rs} + V_{rs} = \delta_{rs}\varepsilon_r \tag{2.85}$$

The BT [Eq. (2.81)] can now be used to determine V and hence to determine the spin-orbitals ϕ_r. By inserting the H of Eq. (2.84) into Eq. (2.81) we obtain

$$0 = \langle 0|[r^+s, H]|0\rangle = (\varepsilon_s - \varepsilon_r)\langle 0|r^+s|0\rangle + \sum_j (V_{jr}\langle 0|j^+s|0\rangle - V_{sj}\langle 0|r^+j|0\rangle)$$
$$+ \sum_{k,l,j} (\langle sk||jl\rangle\langle 0|r^+k^+lj|0\rangle + \langle lk||jr\rangle\langle 0|l^+k^+js|0\rangle) \tag{2.86}$$

where

$$\langle ij||kl\rangle \equiv \langle ij|kl\rangle - \langle ij|lk\rangle \tag{2.87}$$

Since the one-electron density matrix is diagonal for the single-configuration case considered here, we have

$$\langle 0|r^+ s|0\rangle = \delta_{rs}\nu_s \tag{2.88}$$

where ν_s denotes the occupation number for orbital ϕ_s in $|0\rangle$. Because $|0\rangle$ may consist of a linear combination of determinants, the ν_s are not necessarily zero or unity. The Fock potential determined from Eq. (2.86) is then

$$V_{rs}(\nu_r - \nu_s) = \sum_{k,l,j} (\langle sk||jl\rangle\langle 0|r^+k^+lj|0\rangle + \langle lk||jr\rangle\langle 0|l^+k^+js|0\rangle) \tag{2.89}$$

which is only defined from the Brillouin condition when $\nu_r - \nu_s$ is nonzero. Notice that the symmetry of the Fock operator defined in Eq. (2.82) is determined by the symmetry of the above Fock potential. This in turn depends upon the symmetry of the density matrices appearing in Eq. (2.89). As a result, the Fock operator may not have the same symmetry as the full electronic Hamiltonian for specific choice of the reference state $|0\rangle$.

Before discussing various possibilities for how to choose the part of the Fock potential that is not determined from the BT, let us describe the iterative procedure that can be used for obtaining a set of optimized orbitals given any final choice for the form of the full Fock potential. From an initial guess of orbtials, we use Eq. (2.89) together with one of many choices of the remainder of the V to determine a Fock potential. The Fock matrix $\mathsf{F} = \mathsf{h} + \mathsf{V}$ (which is hermitian) is then diagonalized, and a new set of orbitals is determined, which are then used to set up a new Fock potential. This (first-order) process is continued until convergence. The above HF iteration process is nothing but a variant of the commonly used Roothaan SCF procedure (Roothaan, 1951, 1960).

3. Choices of the Nondefined Blocks of the Fock Potential

The part of the Fock potential not defined through the Brillouin condition is often chosen on physical ground [e.g., to have the resultant orbital energies represent ionization potentials and electron affinities (via Koopmans' theorem)] (McWeeney and Sutcliffe, 1976). For a reference state containing a set of occupied spin-orbitals that we denote by α, β, γ, δ and a set of unoccupied spin-orbitals denoted m, n, p, q, the Fock potential in Eq. (2.89) is defined by the BT only between occupied and unoccupied orbitals. From Eq. (2.89) we get

$$V_{m\alpha} = \sum_{\gamma} \langle m\gamma||\alpha\gamma\rangle \tag{2.90}$$

One commonly used choice of the remaining blocks of V results in a Fock potential that can be expressed as

$$V = \sum_{\gamma,r,s} \langle r\gamma || s\gamma \rangle r^+ s \tag{2.91}$$

where r and s run over all spin-orbitals. With this choice, the orbital energies represent (through Koopmans' theorem) the ionization potentials and electron affinities of $|0\rangle$. Of course, other choices of the nondetermined part of V have been made in the literature. For example, the (unoccupied–unoccupied) part of the Fock potential (V_{mn}) has been chosen to correspond to a so-called V^{N-1} potential (Kelly, 1964), thereby making the virtual orbitals more suitable for use in the calculation of excitation energies.

Calculations such as the one discussed above do not involve imposed symmetry restrictions on the reference wavefunction. Hence this approach is referred to as the unrestricted Hartree–Fock (UHF) method. When symmetry restrictions are imposed upon the reference wavefunction the resulting calculation is denoted a restricted Hartree–Fock (RHF) calculation. When the simplest RHF type calculation is carried out for a closed-shell reference state (i.e., one having doubly occupied orbitals), the nondefined part of the Fock potential (the occupied–occupied) and (empty–empty) part is often chosen to have the same form as the (occupied–empty) part defined from the BT. We then would obtain for the entire Fock potential

$$V = \sum_{\substack{\gamma \\ r,s}} (2\langle r\gamma | s\gamma \rangle - \langle r\gamma | \gamma s \rangle)(r_\alpha^+ s_\alpha + r_\beta^+ s_\beta) \tag{2.92}$$

where the indices r, s, and γ refer to orbital indices and the subscripts α, β denote the electron spin m_s component. The orbital energies ε_r then correspond to approximate ionization energies. For a state that has some doubly occupied and some partially filled orbitals, the choice of the nondefined blocks of the Fock potential is less obvious. The BT defines the blocks that connect (occupied–partly occupied), (occupied–empty), and (partly occupied–empty) orbitals. The (occupied–occupied), (partly occupied–partly occupied), and (empty–empty) blocks of the Fock potential are not defined through the BT and many choices have been suggested. One common feature of any of these choices is that the sets of orbitals one obtains in a converged calculation using any arbitrary choice of the nondefined Fock matrix blocks would represent the same SP on the energy hypersurface. The physical interpretations of the orbital energies do, of course, depend on the actual choices made for these "diagonal blocks" of V. For this reason, much work has been devoted to finding particular choices of diagonal blocks that are optimal for particular physical situations. It is not our intention to provide a lengthy discussion of the merits and weaknesses of numerous such methods. Rather, we merely wish to stress that the undetermined blocks of V represent a certain freedom or flexibility that can be exploited to generate orbitals whose orbital energies have some approximate physical meaning.

4. Practical Considerations

Although it is not obvious from Eq (2.89) that a two-electron integral transformation is not required to set up the Fock potential matrix for a general reference state, it becomes clear upon actually working out the matrix elements for a particular case. For example, for either a spin-unrestricted reference state or a closed-shell reference state, the Fock potentials of Eqs. (2.91) and (2.92), respectively, are seen to involve only a two-index transformation [e.g., sum over γ in Eq. (2.92)].

From the above discussion it should be clear that the first-order procedures based upon using the Brillouin condition to define V suffer from some drawbacks. They involve arbitrary choices of certain elements of V (this is related to the invariance of $|0\rangle$ under certain orbital rotations). They are not quadratically convergent and may thus suffer from convergency difficulties. On the other hand, the freedom in choosing elements of V (including the diagonal blocks) is useful when one wishes to cause the resultant orbital energies to have certain physical interpretations (e.g., Koopmans' theorem of ionization energies or excitation energies). The $\exp(i\lambda)$ approach to HF orbital optimization is quadratically convergent but contains no orbital energies for use in physical interpretation. It avoids the problems related to arbitrary choices by simply eliminating from the orbital optimization operator space those operators $(r^+s - s^+r)$ that are redundant and that therefore have no effect on the energy to be extremized.

D. CONFIGURATION INTERACTION METHOD

1. Connection with Second-Order MCSCF Theory

Next we consider the optimization of the total energy when orbital relaxation is not explicitly accounted for in the calculation. The optimization of the total energy may then be carried out either in terms of the configuration expansion coefficients C_{g0} of Eq. (2.4) or in terms of the parameters P of Eq (2.9). Let us consider initially the optimization of the total energy when the configuration expansion coefficients of Eq. (2.4) are used as linear variational parameters. The total energy then becomes

$$E(C_{10}, C_{20}, \ldots) = \sum_{g,g'} C_{g'0}C_{g0}\langle\phi_{g'}|H|\phi_g\rangle \Big/ \sum_g |C_{g0}|^2 \qquad (2.93)$$

where we have assumed that the configuration basis states $|\phi_g\rangle$ are orthonormal. Because the energy function contains no higher than quadratic terms in the C_{g0}, determination of stationary points of the energy function

$$\delta E(C_{10}, C_{20}, \ldots) = 0 \tag{2.94}$$

leads to a set of eigenvalue equations in the configuration expansion coefficients

$$\mathsf{H}\mathbf{C}_0 = E\mathbf{C}_0 \tag{2.95}$$

where H is the matrix representative of the Hamiltonian

$$H_{gg'} = \langle\phi_g|H|\phi_{g'}\rangle \tag{2.96}$$

and the eigenvector

$$\mathbf{C}_0 = \{C_{10}C_{20} \cdots C_{g0}\} \tag{2.97}$$

determines the values of the set of parameters at the SP, where the value of E is E_0. In fact, the same eigenvalue equation, Eq. (2.95), can be used to determine all extrema of the energy within a given configuration space because the energy function contains no more than quadratic terms in $\mathbf{C}$. Equation (2.95) is referred to as the CI eigenvalue equation.

The optimization of the total energy might alternatively be expressed in terms of the variation parameters P [in $\exp(iS)$]. The energy function $E(S)$ would not be quadratic in these parameters P but would contain cubic, quartic, etc. terms in P. An explicit solution from which to determine a SP of the energy function when this unitary $\exp(iS)$ operator is used is very difficult to establish; hence an iterative procedure is required to determine SPs of the energy hypersurface. One iterative scheme that is quadratically convergent is obtained if the terms that refer to the orbital optimization [$\exp(i\lambda)$] are neglected in the MCSCF derivation performed in Section B. The second-order Eq. (2.33) then would read

$$\mathbf{V} = \mathsf{A}_{22}\mathbf{P} \tag{2.98}$$

where A_{22} is defined in Eq. (2.29) as

$$(A_{22})_{mn} = \langle m|H|n\rangle - \delta_{mn}\langle 0|H|0\rangle \tag{2.99}$$

$$\mathbf{V}_n = \langle 0|[R_n, H]|0\rangle = \langle n|H|0\rangle \tag{2.100}$$

and the indices n, m are different from 0. The iterative procedure may be described as follows. For an initial set of configurations ($|0\rangle, |n\rangle$) the matrices A_{22} and $\mathbf{V}$ can be formed. The matrix $\mathbf{P}$ then is determined from Eq. (2.98), and Eq. (2.14) is used to obtain a transformed set of states [one can use alternatively Eqs. (2.15) and (2.16)], and the whole process is repeated until convergence is obtained. The state $|0\rangle$ that would be determined in this iterative procedure would, of course, be the same as that obtained by solving the CI eigenvalue problem.

In the derivation of Section B we considered the energy function to depend on both orbital variation parameters and the configuration expansion coefficients. By freezing the orbital variation parameters, we prohibit orbital relaxation effects from being considered explicitly. To obtain with a CI calculation, which does not permit such orbital relaxation, the same quality as in an MCSCF calculation would require the inclusion of many more configurations, whose purpose would be to compensate for the neglect of explicit orbital relaxation. These additional functions would include a large number of singly excited configurations, but some double, triple, etc. excited configurations would also be needed to fully compensate. If all configurations arising from a given orbital basis were included in a CI calculation (full CI), the need for considering orbital relaxation effects explicitly would, of course, not be present because all orbital variation parameters (λ) would then be redundant variables. However, the number of configurations required to perform a full CI calculation is usually prohibitively large even for systems of modest size. Because CI expansions converge very slowly (as a function of the dimension of the CI secular problem) and the requisite computer time increases very rapidly as more and more configurations are included, efforts must be made to optimize the convergence of a CI calculation by facing two major problems. First, we must make a reasonable choice of orbitals to use in the calculation, and second, the configurations that are to be included in the calculation must be picked by some physically motivated procedure.

2. Choice of Orbitals for Use in CI

The most commonly used set of spin-orbitals for setting up a CI matrix eigenvalue problem is the set of orbitals obtained in a RHF calculation. These orbitals form a particularly convenient set in the sense that they fulfill the BT (i.e., there are no matrix elements connecting the HF ground state and singly excited configurations). However, these orbitals are not especially well suited for use in the CI problem if one desires a reasonably short CI expansion to give high precision. One major problem with the HF orbitals comes from the fact that the electrons in the virtual canonical HF orbitals "feel" an N-electron potential and not an $N-1$ electron potential, as would be physically more proper.

One partial solution to this problem is to use a set of orbitals obtained in a MCSCF calculation for setting up a CI matrix problem whose dimension is much larger than that of the MCSCF calculation. The orbitals would then be relaxed with respect to the configurations included in the MCSCF calculation. Since this MCSCF function normally would include the dominant configurations, a very large fraction of the orbital relaxation effects would would thus explictly be accounted for in the following CI calculations.

Another set of orbitals that has been used as a basis for CI calculations is the so-called iterative natural orbitals (INO) (Bender and Davidson, 1967), which are obtained in the following manner: From a limited number of configurations (the same in all iterations) a reference state $|0\rangle$ is determined by the CI procedure. This reference state is then used to set up the one-electron density matrix $\langle 0|r^+s|0\rangle$, which upon diagonalization gives a set of "natural orbitals." These orbitals are then used for setting up a new CI problem, a new reference state $|0\rangle$ is then determined, and the procedure is continued until a self-consistent set of natural orbitals is determined. Clearly, the INOs are not identical to the MCSCF orbitals discussed earlier. The former are obtained by diagonalizing the first-order density matrix, whereas the MCSCF orbitals are determined by minimizing the electronic energy. The use of INOs in CI calculations is motivated by Löwdin's (Löwdin, 1955) analysis, which showed that such orbitals result in the most compact configuration expansion of $|0\rangle$ (i.e., the fewest configurations being required to generate a wavefunction of a given overlap with the true wavefunction).

The choice of configurations to include in an INO calculation requires particular attention. If the configuration list only includes configurations that are doubly excited with respect to each other, any set of orbitals would be natural orbitals. To make the natural orbital concept useful, the list of configurations has to contain configurations that are singly excited with respect to each other. For example, for the ground state of the beryllium atom, a natural choice of configurations in an MCSCF calculation would be $1s^2 2s^2$ and $1s^2 2p^2$. In the INO calculation, the configuration list would further have to include $1s^2 2sns$ and $1s^2 2pnp$, $n = 3, 4, 5, \ldots$. These configurations would then, to a certain degree, simulate the orbital optimization parameters κ_{ns2s} and κ_{np2p} contained in the MCSCF calculation.

3. Selection of Configurations

Let us now move on to discuss some basic ideas (Schaefer and Miller, 1977, Chapter 6) behind selecting the number of configurations to be included in the CI calculation. With a well-chosen set of orbitals, it is thought that a very small fraction of all possible configurations gives the most important contributions to the total energy. Estimates of the importance of the individual configurations may be obtained from a perturbation theory analysis of the CI secular problem [Eq. (2.95)]. An order analysis based upon Rayleigh–Schrödinger perturbation theory (RSPT) shows immediately the order in which any particular class of configurations enters into the wavefunction. For example, for a set of HF orbitals of a closed-shell system, only the doubly excited configurations contribute to the first-order wavefunction (see Section 3.F). Estimation of the coefficients of the individual configurations through perturbation theory may then be used to select the important configurations by specifying a certain tolerance for the coefficient (or the energy contribution) below which the configurations are not included. For cases in which several configurations are very important to the description of the system, these configurations may be used to form a so-called reference space whose coupling with other configurations can then be estimated through perturbation theory. Another approach is based on performing a series of $(n + 1)$-dimensional CI calculations among the n-dimensional reference space and a sequence of configurations that are obtained as low-order excitations out of these reference functions. The criterion for rejecting configurations tested in this manner usually has to do with the energy lowering of one or more of the n reference-state energies caused by the "added configuration" (Buenker and Peyerimhoff, 1974).

4. Treating Large CI Matrices—Direct Methods

When any such preselection of configurations has been performed, one is often faced with the problem that 10–300,000 configurations have to be included in the final CI calculation. Conventional matrix diagonalization routines such as the one used in the Householder algorithm, which modifies the elements of the matrix as it proceeds, cannot be used to determine the eigenvalues and eigenvectors of the CI matrix. For this reason, specialized approaches have been developed (Schaefer and Miller, 1977, Chapters 7 and 8) to determine a few selected roots (usually the lowest) of such very large CI matrices. One very important feature of these methods is that they do not entail modification of the CI matrix while determining a particular root. To clarify this point, we describe two such techniques, which are referred to as the power method and the perturbation theory method. Although much more efficient approaches have become available, we have chosen to discuss these techniques because they stress, in a simple manner, the basic principles underlying the direct determination of particular eigenstates. In the power method one considers a sequence of operations of the Hamiltonian matrix on an, in principle, arbitrary initial guess of the state vector $\mathbf{C}^0$:

$$\{\mathbf{C}^0, \mathsf{H}\mathbf{C}^0, \mathsf{H}^2\mathbf{C}^0, \ldots, \mathsf{H}^n\mathbf{C}^0\} \tag{2.101}$$

The information content of the converged vector $\lim_{n\to\infty} \mathsf{H}^n\mathbf{C}^0$ can be understood by expanding the vector $\mathbf{C}^0$ in terms of the (unknown) exact eigenvectors $\mathbf{C}_i$ of the Hamiltonian matrix

$$\mathsf{H}\mathbf{C}_i = E_i\mathbf{C}_i \tag{2.102}$$

$$\mathbf{C}^0 = \sum_n a_n\mathbf{C}_n, \qquad a_n = \langle \mathbf{C}^0 | \mathbf{C}_n \rangle \tag{2.103}$$

By assuming that the eigenvalues of H are ordered such that

$$|E_0| \geq |E_1| \geq |E_2| \geq \cdots \geq 0 \tag{2.104}$$

we obtain the formal result

$$\mathsf{H}^n\mathbf{C}^0 = E_0^n\left\{a_0\mathbf{C}_0 + \sum_{j=1} a_j\left(\frac{E_j}{E_0}\right)^n \mathbf{C}_j\right\} \tag{2.105}$$

which, because $|E_j/E_0| < 1$, reduces for large n to

$$\mathsf{H}^n\mathbf{C}^0 = a_0E_0^n\mathbf{C}_0 \tag{2.106}$$

Of course, to arrange the energy ordering assumed above, one might have to subtract from all diagonal elements of H a constant that depends on the largest positive diagonal H_{ii} element. This constant would then be added back onto the resultant E_0 value to obtain the true lowest desired eigenvalue. Hence we see that, for large enough n, the vectors $\mathsf{H}^{n+1}\mathbf{C}^0$ and $\mathsf{H}^n\mathbf{C}^0$ should be proportional, with their proportionality constant equal to E_0, and $\mathbf{C}_0$ should be the eigenvector of the Hamiltonian matrix having the largest eigenvalue E_0. Notice from Eq. (2.106) that the norm of $\mathsf{H}^n\mathbf{C}^0$ grows with n; therefore, normalization of the eigenvector $\mathbf{C}^0$ may be required during the above iterative scheme. E_0 and $\mathbf{C}_0$ are obtained without ever modifying the elements of matrix H; only simple row-by-row multiplication of H with a vector is involved. In fact, as we show below, one can even circumvent the explicit reference to elements of H by using integral-driven matrix multiplication techniques. Such steps become advantageous when one must avoid having to read through the integrals many times. The convergency rate of the power method is governed by the ratio E_1/E_0 and by the choice of $\mathbf{C}^0$. An inappropriate choice of the initial state vector $\mathbf{C}^0$ may lead to slow convergence (e.g., if a_0 vanishes, the power method, in principle, cannot converge to $\mathbf{C}_0$). Once one has obtained the desired E_0 and $\mathbf{C}_0$, the next eigenvalue of H can be found by employing $\mathsf{H} + |E_0|\,|\mathbf{C}_0\rangle\langle\mathbf{C}_0|$ instead of H in the next application of the power method. The lowest root of this

$$(\mathsf{H} + |E_0|\,|\mathbf{C}_0\rangle\langle\mathbf{C}_0|)$$

matrix should then be E_1.

The power method as outlined above is not very widely used in large-scale CI calculations because it is not usually very rapidly convergent. In contrast, variants of the perturbation method described below do constitute viable approaches to finding eigenvalues of large CI matrices. In the basic perturbation method one introduces a decomposition of the full CI Hamiltonian matrix

$$\mathsf{H} = \mathsf{H}^0 + \mathsf{V} \tag{2.107}$$

In what follows, we make the simplest possible choice of H^0; we take it to be the diagonal part of H. Another choice of H^0 that has been widely used. (Davidson, 1975) involves taking H^0 to be a certain small subblock of H (with H_{kk} elements filling the remaining diagonal entries of H^0), which involves the Hamiltonian matrix elements of the most dominant configurations in the desired eigenvector. Given a choice of H^0, the CI secular problem becomes

$$(\mathsf{H}^0 - E)\mathbf{C} = -\mathsf{V}\mathbf{C} \tag{2.108}$$

By iterating on this equation according to the prescription

$$\mathbf{C}^{(n)} = (E - \mathsf{H}^0)^{-1}\mathsf{V}\mathbf{C}^{(n-1)} \tag{2.109}$$

one generates successively higher approximations to the desired $\mathbf{C}$ vector. Corrections to the eigenvalue E are achieved at each iteration by premultiplying Eq. (2.108) on the left by the transpose of $\mathbf{C}^0$ to yield

$$(E^0 - E)(\mathbf{C}^0)^{\mathrm{T}}\mathbf{C} = -(\mathbf{C}^0)^{\mathrm{T}}\mathsf{V}\mathbf{C} \tag{2.110}$$

Initial estimates $\mathbf{C}^0$ and E^0 must, of course, be made consistent with the choice of H^0. For the diagonal choice of H^0, $\mathbf{C}^0$ would correspond to a unit vector $\mathbf{C}^0 = (1, 0, 0, \ldots, 0)$ and E^0 to the diagonal element of H ($E^0 = H_{11}$). If H^0 were taken to be a small subblock of a very large H matrix, Eq. (2.108) could still be solved perturbatively since the dimension of the matrix $(\mathsf{H}^0 - E)$ to be inverted would not be large. The iterative scheme contained in Eqs. (2.109) and (2.110) generates successively higher-order corrections to the desired energy and eigenvector.

To demonstrate how such perturbative methods lead to so-called direct CI techniques, let us consider a simple application of Eqs. (2.109) and (2.110) to a CI wavefunction consisting of a dominant HF configuration $|\phi_{\mathrm{HF}}\rangle$ plus all pair excitations of the form $|\phi_\mu^m\rangle = m_\alpha^+ m_\beta^+ \mu_\beta \mu_\alpha |\phi_{\mathrm{HF}}\rangle$. The elements of the V matrix can be easily written in terms of two-electron integrals

$$V_{\mathrm{HF},m\mu} = \langle \mu\mu | mm \rangle \tag{2.111}$$

$$V_{m\mu,n\nu} = \delta_{mn}\langle \nu\nu | \mu\mu \rangle + \delta_{\mu\nu}\langle mm | nn \rangle \tag{2.112}$$

whereas

$$\langle \phi_{\rm HF}|H^0|\phi_{\rm HF}\rangle = E_{\rm HF} \tag{2.113}$$

and $\langle \phi_\mu^m|H^0|\phi_\mu^m\rangle$ is the Hamiltonian expectation value for the doubly excited configuration $|\phi_\mu^m\rangle$. The matrix product $\mathsf{V}\mathbf{C}^{(n-1)}$ appearing in Eq. (2.109) can be written as follows:

$$\sum_{k\nu} V_{{\rm HF},k\nu} C_{k\nu}^{(n-1)} = \sum_{k\nu} \langle \nu\nu|kk\rangle C_{k\nu}^{(n-1)} \tag{2.114}$$

$$\sum_{k\nu} V_{m\mu,k\nu} C_{k\nu}^{(n-1)} = \sum_{\nu} \langle \nu\nu|\mu\mu\rangle C_{m\nu}^{(n-1)} + \sum_{k} \langle mm|kk\rangle C_{k\mu}^{(n-1)} \tag{2.115}$$

Using these results in Eq. (2.109) we obtain an explicit formula for the elements of $\mathbf{C}^{(n)}$:

$$C_{\rm HF}^{(n)} = \sum_{k\nu} (E - E_{\rm HF})^{-1} \langle \nu\nu|kk\rangle C_{k\nu}^{(n-1)} \tag{2.116}$$

$$C_{m\mu}^{(n)} = (E - \langle \phi_\mu^m|H^0|\phi_\mu^m\rangle)^{-1} \left(\sum_{\nu} \langle \nu\nu|\mu\mu\rangle C_{m\nu}^{(n-1)} + \sum_{k} \langle mm|kk\rangle C_{k\mu}^{(n-1)} \right) \tag{2.117}$$

By writing out the elements of V and H^0 in terms of the integrals, we see that the iterative scheme for the evaluation of $\mathbf{C}$ and E can be written entirely in terms of sums over integrals and $\mathbf{C}^{(n-1)}$ and E values from the preceding iteration. This fact allows this perturbation scheme to be programmed on a computer in an integral-driven manner. That is, as the two-electron integrals $\langle ij|kl\rangle$ are brought into the core memory of the computer, all contributions of each successive integral to all of the sums appearing in Eqs. (2.116) and (2.117) can be evaluated, multiplied by appropriate factors, and added to the appropriate expansion coefficients. In this way, the computer is required to read through the (presumably long) list of two-electron integrals only once for each iteration. In this way, one avoids the explicit construction and storage of the Hamiltonian matrix, which may be very large and much larger than the number of two-electron integrals.

Techniques that permit the working numerical equations [e.g., Eqs. (2.116) and (2.117)] to be expressed as sums over explicit two-electron integrals are referred to as integral-driven direct CI methods. The perturbation solution described above is only a simple example of such methods. For more general classes of CI wavefunctions, the expressions for the V matrix elements are more involved. However, the basic structure and philosophy of the direct CI techniques remain as outlined. These techniques have proven to be quite useful in carrying out large-scale CI calculations, and such integral-driven strategies have been used to efficiently implement the graphical unitary group approach (GUGA) for electronic structure calculations (Paldus and Boyle, 1980; Shavitt, 1978; Brooks and Schaefer, 1979).

5. Size Consistency

Thus far, we have concentrated on describing how the CI procedure is used in practical applications and how it can be viewed as relating to the MCSCF method. It is important to realize that even though difficulties having to do with large CI matrices may be overcome, a serious problem remains inherent in nearly all of the above methods. To understand the difficulty, consider how one might perform a calculation of a potential-energy curve for the diatomic Be_2. Assume that a prior calculation on a single beryllium atom indicated that the $2s^2$ and $2p^2$ ^1S configuration should be included in order to describe the electron correlation in beryllium. Then to describe the correlation in Be_2 in a balanced manner (i.e., such as to yield a $2s^2 + 2p^2$ level description of both beryllium atoms upon dissociation), one must include the $2s_A^2 2s_B^2$, $2s_A^2 2p_B^2$, $2p_A^2 2s_B^2$, and $2p_A^2 2p_B^2$ configurations, where A and B label the two beryllium nuclei. Hence, although a double-excitation CI or MCSCF could be employed for Be, one needs to include (certain) quadruple excitation (relative to $2s_A^2 2s_B^2$) for Be_2. Clearly, for more complex molecular clusters one would need to include even higher level excitations (e.g., eightfold for Be_4) to achieve a qualitatively balanced description of the complex and its fragments. This is, of course, essential if one is trying to compute energy changes (bond energies and energies of formation) for chemical reactions. Then one must use a method that yields the same value for the molecular complex energy (e.g., Be_2) when evaluated at large interfragment separation as the sum of the fragment (e.g., two beryllium atoms) energies evaluated separately within the same method. Such methods are said to be size consistent (Pople *et al.*, 1977). The use of a restricted CI or MCSCF wavefunction (e.g., doubly excited for Be_2) could indeed yield a smooth potential-energy curve free of obvious pathological behavior. However, such a wavefunction would preferentially describe the electron correlation in the complex (Be_2) near its equilibrium geometry and would dissociate to yield fragments that are described to a lower correlation level (e.g., the $2s_A^2 2s_B^2$ configuration would dominate).

The size consistency problem may be less significant if an appropriate configuration selection is performed at each geometry on the molecular potential surface, but the problem still remains as to how to efficiently choose configurations that describe equally well an entire potential energy surface. It may in fact be more straightforward to achieve this goal using an MCSCF wavefunction, since the orbital optimization thereby included can make the configuration expansion length short enough to be physically understood and hence correctly chosen. As we discuss in more detail in Chapter 3, this question relating to achieving a balanced description of a molecule and its fragments is important and not easily addressed within a variational framework.

6. Discussion

Because the CI technique has been the most widely used approach to treating electron correlation problems, many advances have been made in matrix storage techniques, two-electron integral transformations, the use of unitary group tools, matrix eigenvalue and eigenvector determinations, and configuration selection processes. We by no means intend to treat these advances here; many of them are reviewed well in Chapters 6–8 of Schaefer and Miller (1977). It is essential that one realize that the monumental development of exactly these same data management methods is what makes it possible to implement not only efficient CI computer programs but also highly efficient MCSCF, HF, coupled-cluster, and Green's function routines. To implement any of the above quantum-chemical methods in a state-of-the-art manner, one must make extensive use of many of the advances in numerical methods and data handling that the scientists who have been instrumental in developing efficient CI programs have made.

PROBLEMS

2.1 Using the one- and two-electron integrals given below, carry out an SCF calculation for the $1\sigma^2\mathrm{HeH}^+$ ground state using a first-order procedure.

1. By expanding the molecular orbitals $\{\phi_k\}$ as linear combinations of atomic orbitals $\{\chi_\mu\}$,

$$\phi_k = \sum_\mu c_{\mu k}\chi_\mu$$

and using the definition of the closed-shell Fock operator given in Eq. (2.92), show that the Fock eigenvalue equation can be written in terms of the atomic orbital basis as

$$\mathbf{Fc} = \mathbf{Sc}\boldsymbol{\varepsilon}$$

where the overlap matrix is

$$S_{\mu\nu} = \langle\mu|\nu\rangle$$

the elements of the Fock matrix are

$$F_{\mu\nu} = \langle\mu|h|\nu\rangle + \sum_{\rho\sigma} P_{\rho\sigma}\{2\langle\mu\rho|\nu\sigma\rangle - \langle\mu\rho|\sigma\nu\rangle\} \tag{A}$$

h is the one-electron operator in the Hamiltonian, and the charge bond order matrix $\mathbf{P}$ is defined as

$$P_{\rho\sigma} = \sum_k c^*_{\rho k}c_{\sigma k}$$

2. Show that the HF total energy for a closed-shell system may be written in terms of integrals over the orthonormal HF orbitals as

$$E(\mathrm{SCF}) = 2\sum_{k}^{\mathrm{occ}} \langle \phi_k|h|\phi_k\rangle + \sum_{kl}^{\mathrm{occ}} \{2\langle kl|kl\rangle - \langle kl|lk\rangle\} + \sum_{\mu>\nu} (Z_\mu Z_\nu/R_{\mu\nu}) \qquad \text{(B)}$$

3. Show that the HF total energy may alternatively be expressed as

$$E(\mathrm{SCF}) = \sum_{k}^{\mathrm{occ}} \{\varepsilon_k + \langle \phi_k|h|\phi_k\rangle\} + \sum_{\mu>\nu}(Z_\mu Z_\nu/R_{\mu\nu}) \qquad \text{(C)}$$

where the $\{\varepsilon_k\}$ refer to the HF orbital energies.

To carry out an SCF calculation on the ground state of HeH^+ at $R =$ 1.4 a.u., the following information is to be used. The orbital exponents of the 1s, Slater orbitals of the He and H are 1.6875 and 1.0, respectively. The atomic integrals required to carry out the HF calculation are (in a.u.)

$$S_{11} = S_{22} = 1.0, \qquad S_{12} = 0.5784$$

$$h_{11} = -2.6442, \quad h_{22} = -1.7201, \quad h_{12} = -1.5113, \quad (h_{ij} \equiv \langle i|h|j\rangle),$$

$$\langle 11|11\rangle = 1.0547, \quad \langle 11|21\rangle = 0.4744, \quad \langle 12|12\rangle = 0.5664,$$

$$\langle 22|11\rangle = 0.2469, \quad \langle 22|21\rangle = 0.3504, \quad \langle 22|22\rangle = 0.6250$$

where 1 refers to $1s_{\mathrm{He}}$ and 2 to $1s_{\mathrm{H}}$. In this and the following problems we shall employ the indices 1 and 2 to label either the molecular orbitals or the atomic orbitals whenever doing so is not confusing. We shall reserve the notation 1σ and 2σ primarily for describing the orbital occupancies arising in the wavefunctions. As an initial guess for the occupied molecular orbital use $\phi_1 \simeq 1s_{\mathrm{He}}$.

4. Form, with this initial guess of the occupied molecular orbital, a 2×2 Fock matrix, using Eq. (A) for $F_{\mu\nu}$.

5. Solve the Fock matrix eigenvalue equations given above to obtain the orbital energies and an improved occupied molecular orbital. In so doing, note that the normalization condition $\langle \phi_1|\phi_1\rangle = 1 = \mathbf{C}_1^{\mathrm{T}}\mathbf{S}\mathbf{C}_1$ gives the needed normalization condition for the expansion coefficients of the ϕ_1 in the atomic orbital basis.

6. Determine the total SCF energy using Eq. (C) at this step of the iterative procedure. When will this energy agree with that obtained by using the alternative expression for $E(\mathrm{SCF})$ given in Eq. (B)?

7. Use the ϕ_1 molecular orbital from question 5 to determine a new Fock matrix.

8. Determine a new set of orbital energies and an improved occupied molecular orbital.

9. Determine from Eq. (C) the SCF total energy at this step of the iterative procedure.

The iterative process may be continued until convergence. As a convergence criterion, assume that the difference between the SCF total energy in two successive iterations must be less than 10^{-6} a.u. Listed below are the HF total energies (in a.u.) obtained during the iterative procedure beyond the two iterations performed above:

$$-2.842151, \quad -2.843221, \quad -2.843393,$$
$$-2.843420, \quad -2.843425, \quad -2.843425$$

10. Show, by comparing the difference between the SCF total energy at one iteration and the converged SCF total energy, that the convergence of the above SCF approach is linear or first order.

11. Is the SCF total energy listed above in each iteration of the SCF procedure an upper bound to the exact ground-state total energy?

The converged self-consistent set of molecular orbitals ϕ_1 and ϕ_2 is

$$\phi_1 = 0.9000\ 1s_{\mathrm{He}} + 0.1584\ 1s_{\mathrm{H}}, \qquad \phi_2 = -0.8324\ 1s_{\mathrm{He}} + 1.2156\ 1s_{\mathrm{H}}$$

12. Show, using the one- and two-electron integrals in the molecular orbital basis,

$$\langle 1|h|1\rangle = -2.6158, \qquad \langle 1|h|2\rangle = 0.1954, \qquad \langle 2|h|2\rangle = -1.3154$$
$$\langle 11|11\rangle = 0.9596, \qquad \langle 11|21\rangle = -0.1954, \qquad \langle 12|12\rangle = 0.6063,$$
$$\langle 12|21\rangle = 0.1261, \qquad \langle 22|21\rangle = -0.0045, \qquad \langle 22|22\rangle = 0.6159$$

that the converged values of the orbital energies are

$$\varepsilon_1 = -1.6562, \qquad \varepsilon_2 = -0.2289$$

13. Does this SCF wavefunction give rise (at $R \to \infty$) to proper dissociation products?

2.2 Now carry out an SCF calculation for the same closed-shell HeH^+ system using a second-order SCF procedure. Some of the integrals used in Problem 2.1 will be useful again here.

1. Show that the one- and two-electron density matrices decouple as follows for a closed-shell reference state:

$$\sum_{\sigma} \langle 0|r_\sigma^+ s_\sigma|0\rangle = \delta_{rs} 2\nu_r, \qquad \sum_{\sigma\sigma'} \langle 0|r_\sigma^+ s_{\sigma'}^+ t_{\sigma'} u_\sigma|0\rangle = (4\delta_{ru}\delta_{st} - 2\delta_{rt}\delta_{su})\nu_r \nu_s$$

where ν_r is the occupation of orbital ϕ_r. That is, if ϕ_r is an occupied orbital $\nu_r = 1$, and if ϕ_r is unoccupied $\nu_r = 0$.

2. Show that the A_{11}, B_{11}, and W matrices of Eqs. (2.29), (2.30), and (2.27), respectively, may be written for a closed-shell system as

$$(A_{11})_{n\beta,m\alpha} = \langle 0|[\beta_\alpha^+ n_\alpha + \beta_\beta^+ m_\beta, H, m_\alpha^+ \alpha_\alpha + m_\beta^+ \alpha_\beta]|0\rangle$$

$$= 2\Big[-h_{\alpha\beta}\delta_{mn} + h_{nm}\delta_{\alpha\beta} + \delta_{\alpha\beta}\sum_\gamma \{2\langle n\gamma|m\gamma\rangle - \langle n\gamma|\gamma m\rangle\}$$

$$-\delta_{mn}\sum_\gamma \{2\langle \alpha\gamma|\beta\gamma\rangle - \langle\alpha\gamma|\gamma\beta\rangle\}$$

$$+ 2\langle n\alpha|\beta m\rangle - \langle n\alpha|m\beta\rangle\Big]$$

$$(B_{11})_{n\beta,m\alpha} = \langle 0|[\beta_\alpha^+ n_\alpha + \beta_\beta^+ n_\beta, H, \alpha_\alpha^+ m_\alpha + \alpha_\beta^+ m_\beta]|0\rangle$$

$$= 2[\langle mn|\beta\alpha\rangle - 2\langle mn|\alpha\beta\rangle]$$

$$W_{\alpha m} = \langle 0|[\alpha_\alpha^+ m_\alpha + \alpha_\beta^+ m_\beta, H]|0\rangle = 2\Big[h_{m\alpha}$$

$$+\sum_\gamma \{2\langle m\gamma|\alpha\gamma\rangle - \langle m\gamma|\gamma\alpha\rangle\}\Big]$$

Again use as the initial guess of the occupied molecular orbital $1s_{\mathrm{He}}$.

3. Given this guess for ϕ_1, determine the virtual or unoccupied molecular orbital ϕ_2 using a Schmidt orthogonalization procedure. The atomic integrals required are given in Problem 2.1.

The second-order SCF procedure requires knowledge of the integrals in the basis of the set of initial orthonormal molecular orbitals (ϕ_1 and ϕ_2 obtained above). The one- and two-electron integrals in this basis are given below (in a.u.):

$$h_{11} = -2.6442, \qquad h_{22} = -1.2870, \qquad h_{12} = 0.0223$$

$$\langle 11|11\rangle = 1.0547, \qquad \langle 11|21\rangle = -0.1663, \qquad \langle 12|12\rangle = 0.5567,$$

$$\langle 22|11\rangle = 0.0765, \qquad \langle 22|21\rangle = 0.0171, \qquad \langle 22|22\rangle = 0.6200$$

where, as before, 1 denotes the occupied and 2 the unoccupied molecular orbital.

4. Determine the SCF total energy that corresponds to this initial guess of molecular orbitals.

5. Determine the A_{11}, B_{11}, and W matrix elements.

6. Determine the $\boldsymbol{\kappa}$ matrix and the unitary matrix $\mathsf{X} = e^{-\boldsymbol{\kappa}}$.

7. Determine the new improved set of orthonomal molecular orbitals resulting from applying X to ϕ_1 and ϕ_2.

The one- and two-electron integrals may now be evaluated in the set of improved molecular orbitals and the iterative procedure thus may be continued until convergence is obtained. The HF total energies obtained during the iterative procedure become

$$-2.80504513, \quad -2.84303574, \quad -2.84342526, \quad -2.84342527$$

8. Show by comparing the difference between the SCF total energies at successive iterations and the converged SCF total energy that the convergence of the above SCF approach is quadratic or second order.

2.3 Given the one- and two-electron integrals in the SCF orbital basis found in Problem 2.1, carry out a two-configuration CI calculation on HeH^+ using the $1\sigma^2$ and $2\sigma^2$ configurations.

1. First obtain expressions for the CI matrix elements $H_{ij}\,(i, j = 1\sigma^2, 2\sigma^2)$ in terms of one- and two-electron integrals.

2. Show that the resultant CI matrix is (ignoring the nuclear repulsion term)

$$\begin{pmatrix} -4.2720 & 0.1261 \\ 0.1261 & -2.0149 \end{pmatrix}$$

3. Obtain the two CI energies and eigenvectors for the above matrix.

4. Show that the lowest-energy CI wavefunction is equivalent to the following two-determinant (single configuration) wavefunction:

$$\tfrac{1}{2}\left[\left|(a^{1/2}\phi_1 + b^{1/2}\phi_2)\alpha(a^{1/2}\phi_1 - b^{1/2}\phi_2)\beta\right| + \left|(a^{1/2}\phi_1 - b^{1/2}\phi_2)\alpha(a^{1/2}\phi_1 + b^{1/2}\phi_2)\beta\right|\right]$$

involving the polarized orbitals $a^{1/2}\phi_1 \pm b^{1/2}\phi_2$, where $a = 0.9984$ and $b = 0.0556$.

2.4 Using the same information as in Problem 2.3, carry out a three-configuration CI calculation on HeH^+ at $R = 1.4$ a.u. using the $1\sigma^2$, $2\sigma^2$, and $1\sigma 2\sigma$ electronic configurations.

1. First express the proper singlet spin-coupled $1\sigma 2\sigma$ configuration as a combination of Slater determinants.

2. Compute all elements of the 3×3 CI matrix.

3. Obtain the eigenenergies and corresponding normalized eigenvectors for this problem.

2.5 Use the perturbative method described in Section D.4 on the CI matrix eigenvalue problem of Problem 2.4 to find the lowest eigenenergy and its corresponding eigenvalues. Use as the initial guess for the eigenvector $\mathbf{C}^0 = (1.0000, 0.0, 0.0)$ and take

$$\mathsf{H}^0 = \begin{pmatrix} -4.2720 & 0 & 0 \\ 0 & -2.0149 & 0 \\ 0 & 0 & -3.1988 \end{pmatrix}$$

and $E^0 = -4.2720$ for the first iteration. Use the energy computed using Eq. (2.110) to start the second iteration, but notice that the $\mathbf{C}^{(1)}$ vector you then obtain is so much different from $\mathbf{C}^{(0)}$ that convergence of the process is not likely. Therefore, average these $\mathbf{C}^{(0)}$ and $\mathbf{C}^{(1)}$ to obtain a new damped $\mathbf{C}^{(1)}$ for use in determining $\mathbf{C}^{(2)}$.

2.6 Perform a one-step second-order multiconfiguration HF calculation on HeH^+, using the minimum Slater basis of Problem 2.1. The multiconfiguration reference state will include the two configurations $1\sigma^2$ and $2\sigma^2$. As an initial guess of orbitals use the set of single-configuration HF orbitals of the principal configuration $1\sigma^2$. The HF orbitals were determined in Problem 2.1, and the one- and two-electron integrals in the HF basis are given there. The initial guess of the configuration state functions (denoted $|0\rangle$ and $|1\rangle$) will be the ones determined in the two-configuration CI calculation given in Problem 2.3.

1. Determine all of the nonvanishing one- and two-electron density matrix elements

$$\langle 0|r^\dagger s|0\rangle, \langle 0|r^\dagger s^\dagger tu|0\rangle$$

and the nonvanishing one- and two-electron transition density matrix elements

$$\langle 1|r^\dagger s|0\rangle, \langle 1|r^\dagger s^\dagger tu|0\rangle$$

2. Determine the $\mathbf{V}$, $\mathbf{W}$, A, and B matrix elements.
3. Determine the $\boldsymbol{\kappa}$ and the $\mathbf{P}$ matrix elements via the one-step second-order MCSCF method.
4. Determine the transformed set of orbitals and states ($|0\rangle$ and $|1\rangle$).
5. Discuss whether the orbitals and states obtained after the first iteration of the one-step second-order MCSCF procedure (question 4) differ from the orbitals and states that would be obtained after the first iteration of the two-step second order MCSCF procedure. If they differ, describe how they would be obtained in the two-step procedure.

From the orbitals and states obtained in question 4 new one- and two-electron integrals and one- and two-electron density and transition density matrix elements may now be evaluated, and the iterative procedure thus continued. The multiconfigurational HF total energies obtained during this iterative procedure are

$$-2.85044942, \quad -2.85066435, \quad -2.85066436$$

6. Show by comparing the difference between the MCSCF total energies at each iteration and the converged MCSCF total energy that the convergence rate of the used MCSCF approach is second order.
7. How must the converged MCSCF ground-state total energy compare with the ground-state total energy obtained in the full CI calculation?

8. Write a priori the ground-state total energy that would be obtained if we used the three configurations $1\sigma^2$, $2\sigma^2$, and $1\sigma^1 2\sigma^1$ in an MCSCF calculation.

9. Write a priori the ground-state total energy that would be obtained from a converged two-configuration MCSCF calculation that used the $|1\sigma^2\rangle$ and $|1\sigma 2\sigma\rangle$ configurations.

2.7 Consider n HeH^+ molecular ions, which do not interact because they are infinitely far from one another.

1. Write the electronic Hamiltonian for this system in a basis consisting of orthonormal orbitals that are localized on each of the HeH^+ molecules. Retain only those contributions that are nonzero. In so doing, describe each HeH^+ molecule with a bonding and antibonding SCF orbital pair.

2. Show that a CI calculation that includes the HF ground-state wavefunction consisting of the antisymmetrized product of orbitals localized on the n ions having $1\sigma^2$ occupancy, and all doubly excited configurations leads to the following CI matrix of dimension $n + 1$:

$$\begin{pmatrix} nE_{\text{HF}} & B & B & B & \cdots & B \\ B & C & 0 & 0 & & \\ B & 0 & C & 0 & & \\ B & 0 & 0 & C & & \\ \vdots & & & & \ddots & \\ B & & & & & C \end{pmatrix}$$

where

$$C = E^*_{\text{HF}} - E_{\text{HF}} + nE_{\text{HF}}, \qquad B = \langle 11|22\rangle = \langle 1\sigma^2|H|2\sigma^2\rangle$$
$$E_{\text{HF}} = 2h_{11} + \langle 11|11\rangle = \langle 1\sigma^2|H|1\sigma^2\rangle,$$
$$E^*_{\text{HF}} = 2h_{22} + \langle 22|22\rangle = \langle 2\sigma^2|H|2\sigma^2\rangle$$

As in other problems, 1 and 2 denote the bonding and antibonding SCF molecular orbitals, respectively, for an isolated HeH^+ molecule.

3. Show that the correlation energy for n infinitely separated HeH^+ molecules is

$$E_{\text{corr}} = \frac{-E_{\text{HF}} + E^*_{\text{HF}}}{2} - \left[\frac{(-E_{\text{HF}} + E^*_{\text{HF}})^2}{4} + n\langle 11|22\rangle^2\right]^{1/2}$$

4. Use the HeH^+ SCF orbitals and results from Problems 2.1 and 2.3 to evaluate for $n = 2$, 4, 10, 100, and 1000 the correlation energy obtained for n infinitely separated HeH^+ molecules. Show that the correlation energy increases as $n^{1/2}$ when n becomes large. How would the correlation energy increase in a size-consistent model?

Bartlett and Purvis (1981) have found that for H_2 and He the percentage errors caused by size inconsistency in double-excitation CI calculation vary as follows:

n	Error H_2 (%)	Error He (%)
2	1.5	0.8
4	4.8	2.4
10	12.3	6.5
100	48.0	34.8
1000	79.1	70.8

5. Argue why the two-basis function HeH^+ problem is likely to underestimate the non-size-consistent contributions when compared with results obtained in more accurate calculations on HeH^+.

SOLUTIONS

2.1

1. $V_{ij} = \sum_{\gamma}(2\langle i\gamma|j\gamma\rangle - \langle i\gamma|\gamma j\rangle)$

Let

$$\phi_i = \sum_{\mu} C_{\mu i}\chi_{\mu}, \qquad \phi_{\gamma} = \sum_{\nu} C_{\nu\gamma}\chi_{\nu}$$

Then

$$V_{ij} = \sum_{\mu'\mu\nu\nu'} (C_{\nu\gamma}C_{\nu'\gamma})(C_{\mu i}C_{\mu' j})(2\langle\mu\nu|\mu'\nu'\rangle - \langle\mu\nu|\nu'\mu'\rangle)$$
$$= \sum_{\mu\mu'} C_{\mu i}C_{\mu' j}V_{\mu\mu'}$$

where

$$V_{\mu\mu'} = \sum_{\nu\nu'} P_{\nu\nu'}(2\langle\mu\nu|\mu'\nu'\rangle - \langle\mu\nu|\nu'\mu'\rangle), \qquad P_{\nu\nu'} = \sum_{\gamma} C_{\nu\gamma}C_{\nu'\gamma}$$

Likewise

$$\langle\phi_i| - \tfrac{1}{2}\nabla^2 - \sum_A (Z_A/|\mathbf{r} - \mathbf{R}_A|)\,|\phi_j\rangle \equiv h_{ij} = \sum_{\mu\mu'} C_{\mu i}C_{\mu' j}h_{\mu\mu'}$$
$$h_{\mu\mu'} = \langle\chi_\mu| - \tfrac{1}{2}\nabla^2 - \sum_A (Z_A/|\mathbf{r} - \mathbf{R}_A|)|\chi_{\mu'}\rangle$$

As a result $F\phi_i = \varepsilon_i\phi_i$ can, by expanding ϕ_i as above, be expressed as

$$h_{ij} + V_{ij} = \delta_{ij}\varepsilon_i = \sum_{\mu\mu'} C_{\mu i}C_{\mu' j}(h_{\mu\mu'} + V_{\mu\mu'})$$

Then using

$$\langle \phi_i | \phi_j \rangle = \delta_{ij} = \sum_{\mu\mu'} C_{\mu i} S_{\mu\mu'} C_{\mu' j}$$

we have

$$\sum_{\mu\mu'} C_{\mu i} [\varepsilon_j S_{\mu\mu'} - h_{\mu\mu'} - V_{\mu\mu'}] C_{\mu' j} = 0, \qquad \text{for all} \quad i, j$$

This can only be true if

$$\sum_{\mu'} (h_{\mu\mu'} + V_{\mu\mu'} - \varepsilon_j S_{\mu\mu'}) C_{\mu' j} = 0$$

This is $\mathbf{FC} = \mathbf{SC}\boldsymbol{\varepsilon}$.

2. The Slater–Condon rules tell us that the Hamiltonian expectation value for a single Slater determinant in which spin orbitals $\phi_1, \ldots, \phi_N$ are occupied is

$$E = \sum_{k=1}^{N} \langle \phi_k | -\tfrac{1}{2}\nabla^2 - \sum_A (Z_A/|\mathbf{r} - \mathbf{R}_A|) | \phi_k \rangle + \frac{1}{2} \sum_{k,l} [\langle kl|kl \rangle - \langle kl|lk \rangle]$$

For a closed-shell system the orbitals are doubly occupied and therefore $\phi_1 = \bar{\phi}_1 \alpha$, $\phi_2 = \bar{\phi}_1 \beta$, $\phi_3 = \bar{\phi}_2 \alpha$, $\phi_4 = \bar{\phi}_2 \beta$, etc., where $\bar{\phi}_1$, $\bar{\phi}_2$, etc. label the occupied orbitals (not spin-orbitals). Hence by carrying out the spin integration in the above energy expression and using the fact that each orbital is doubly occupied, we obtain

$$E = 2 \sum_{k}^{\text{occ}} \langle \phi_k | h | \phi_k \rangle + \sum_{kl}^{\text{occ}} \{ 2\langle kl|kl \rangle - \langle kl|lk \rangle \}$$

where labels now refer to orbital index. The term $\sum_{\mu > \nu} (Z_\mu Z_\nu / R_{\mu\nu})$ must then be added on to obtain the total energy (including nuclear repulsion).

3. If the occupied orbitals ϕ_k obey $F\phi_k = \varepsilon_k \phi_k$ then the above expression for E can be rearranged to give

$$E = \sum_{k}^{\text{occ}} \{ \langle \phi_k | h | \phi_k \rangle + \sum_{l}^{\text{occ}} [2\langle kl|kl \rangle - \langle kl|lk \rangle] \} + \sum_{k}^{\text{occ}} \langle \phi_k | h | \phi_k \rangle$$

The first two terms in this expression can be recognized as $\langle \phi_k | F | \phi_k \rangle$, where F is the closed-shell Fock operator whose potential is defined in Eq. (2.92). Hence

$$E = \sum_{k}^{\text{occ}} \langle \phi_k | F | \phi_k \rangle + \sum_{k}^{\text{occ}} \langle \phi_k | h | \phi_k \rangle$$

4. $\mathbf{P} = \begin{pmatrix} 1.0 & 0.0 \\ 0.0 & 0.0 \end{pmatrix}$, $\quad \mathbf{F} = \begin{pmatrix} -1.5895 & -1.0369 \\ -1.0369 & -0.8342 \end{pmatrix}$

5. $\varepsilon_1 = -1.6048$, $\varepsilon_2 = -0.2348$, $\phi_1 = 0.9194\ 1s_{\text{He}} + 0.1296\ 1s_{\text{H}}$

6. $E_{\text{SCF}} = -2.8005$. The two expressions will agree only upon convergence of the SCF when $F\phi_i = \varepsilon_i\phi_i$, which was assumed in writing the expression for E containing the orbital energies.

7. $\mathsf{P} = \begin{pmatrix} 0.8453 & 0.1192 \\ 0.1192 & 0.0168 \end{pmatrix}$, $\mathsf{F} = \begin{pmatrix} -1.6246 & -1.0836 \\ -1.0836 & -0.8772 \end{pmatrix}$

8. $\varepsilon_1 = -1.6469$, $\varepsilon_2 = -0.2289$, $\phi_1 = 0.9032\ 1s_{\text{He}} + 0.1537\ 1s_{\text{H}}$

9. $E_{\text{SCF}} = -2.8356$

10.

$E_{\text{SCF}} - E_{\text{SCF}}^{\text{conv}}$	$(E_{\text{SCF}} - E_{\text{SCF}}^{\text{conv}})^2$
0.001274	0.000002
0.000204	0.000000
0.000032	
0.000005	
0.000000	

Second-order convergence requires that the error in the $(n + 1)$th iteration is the square of the error in the nth iteration, In the first iteration above the error is 0.001274; thus in the next iteration the error should be $(0.001274)^2 = 0.0000016$ if we used a second-order procedure. Since the second iteration's error is 0.000204, the convergency of the above SCF procedure is linear rather than quadratic.

11. The *converged* SCF total energy calculated from Eq. (C) is an upper bound to the ground-state energy, whereas the SCF total energy from Eq. (C) during the iterative procedure is not a bound. It is only at convergence that the expectation value of the Hamiltonian for the HF determinant is given by Eq. (C).

12. The SCF orbital energies are determined to be

$$\varepsilon_k = \langle k|h|k\rangle + \sum_l \{2\langle kl|kl\rangle - \langle kl|lk\rangle\}$$

from which the orbital energies follow straightforwardly.

13. Yes, the $1\sigma^2$ configuration does dissociate properly because at $R \to \infty$, the lowest-energy state is He + H^+, which also has a $1\sigma^2$ orbital occupancy.

2.2

1. Since ϕ_r and ϕ_s are either occupied or unoccupied $\sum_\sigma \langle 0|r_\sigma^+ s_\sigma|0\rangle$ vanishes unless both ϕ_r and ϕ_s are in $|0\rangle$. Hence $\sum_\sigma \langle 0|r_\sigma^+ s_\sigma|0\rangle = \delta_{rs}2v_r$. Likewise, in $\sum_{\sigma\sigma'} \langle 0|r_\sigma^+ s_{\sigma'}^+ t_{\sigma'} u_\sigma|0\rangle$ all four spin-orbitals must be in $|0\rangle$. Then

$$\begin{aligned}\langle 0|r_\sigma^+ s_{\sigma'}^+ t_{\sigma'} u_\sigma|0\rangle &= \delta_{st}\langle 0|r_\sigma^+ u_\sigma|0\rangle - \langle 0|r_\sigma^+ t_{\sigma'} s_{\sigma'}^+ u_\sigma|0\rangle \\ &= \delta_{st}\delta_{ru} - \delta_{su}\delta_{\sigma\sigma'}\langle 0|r_\sigma^+ t_{\sigma'}|0\rangle + \langle 0|r_\sigma^+ t_{\sigma'} u_\sigma s_{\sigma'}^+|0\rangle \\ &= \delta_{st}\delta_{ru} - \delta_{su}\delta_{rt}\delta_{\sigma\sigma'}\end{aligned}$$

where all orbitals are occupied. Clearly $\sum_{\sigma\sigma'}\langle 0|r_\sigma^+ s_{\sigma'}^+ t_{\sigma'} u_\sigma|0\rangle$ vanishes when one or more of the four orbitals are unoccupied, and its equals $4\delta_{st}\delta_{ru} - 2\delta_{su}\delta_{rt}$ when all four orbitals are occupied.

2. Equation (2.42), when combined with the results of question 1, give

$$(A_{11})_{n\beta,m\alpha} = 2\delta_{\alpha\beta}h_{nm} + 0 - \delta_{nm}2h_{\alpha\beta} - 0 - 0 - \delta_{nm}\sum_\gamma[\langle\alpha\gamma|\beta\gamma\rangle 4 - 2\langle\alpha\gamma|\gamma\beta\rangle]$$
$$- 0 - 0 - \sum_\gamma\langle n\gamma|\gamma m\rangle\delta_{\alpha\beta}2 + 4\langle n\alpha|\beta m\rangle$$
$$+ \delta_{\alpha\beta}\sum_\gamma\langle n\gamma|m\gamma\rangle 4 - 2\langle n\alpha|m\beta\rangle + 0 + 0$$

$$(B_{11})_{n\beta,m\alpha} = 0 + 0 - 0 - 0 - 0 - 0 - 0 - 4\langle mn|\alpha\beta\rangle + 2\langle mn|\beta\alpha\rangle$$
$$+ 0 + 0 + 0 + 0$$

Using Eq. (2.41) we find

$$W_{\alpha m} = 2h_{m\alpha} - 0 - 0 + \sum_\gamma[4\langle m\gamma|\alpha\gamma\rangle - 2\langle m\gamma|\gamma\alpha\rangle]$$

3. $\phi_1 = 1s_{\mathrm{He}}$

$\phi_2 = 1s_{\mathrm{H}} - \langle 1s_{\mathrm{H}}|1s_{\mathrm{He}}\rangle 1s_{\mathrm{He}} = 1s_{\mathrm{H}} - 0.5784\ 1s_{\mathrm{He}}$

Normalizing ϕ_2 requires that we divide by the square root of $1 + \langle 1s_{\mathrm{H}}|1s_{\mathrm{He}}\rangle^2 - 2\langle 1s_{\mathrm{H}}|1s_{\mathrm{He}}\rangle^2$ to obtain

$$\phi_2 = 1.2259\ 1s_{\mathrm{H}} - 0.7091\ 1s_{\mathrm{He}}$$

The coefficient matrix C whose elements $C_{\mu k}$ are the orbital expansion coefficients then becomes

$$\mathsf{C} = \begin{pmatrix} 1.0 & -0.7091 \\ 0.0 & 1.2259 \end{pmatrix}$$

4. -2.8050

5. The excitation from molecular orbital 1 to 2 is nonredundant. The relevant matrices thus become one dimensional:

$$A_{11} = 1.8713, \qquad B_{11} = -0.1530, \qquad W = -0.2880$$

6. $\boldsymbol{\kappa} = \begin{pmatrix} 0.0 & 0.1423 \\ -0.1423 & 0.0 \end{pmatrix}, \qquad \mathsf{X} = \begin{pmatrix} 0.9899 & -0.1418 \\ 0.1418 & 0.9899 \end{pmatrix}$

7. $\mathsf{C} = \begin{pmatrix} 0.8893 & -0.8437 \\ 0.1739 & 1.2135 \end{pmatrix}$

8.

$E_{SCF} - E_{SCF}^{conv}$	$(E_{SCF} - E_{SCF}^{conv})^2$
0.03838013	0.00147303
0.00038952	0.00000015
0.00000001	

The error in the $(n + 1)$th iteration is the square of the error in the nth iteration.

2.3

1. $\langle 1\sigma^2|H|1\sigma^2\rangle = 2h_{11} + \langle 1\,1|1\,1\rangle$, $\langle 2\sigma^2|H|2\sigma^2\rangle = 2h_{22} + \langle 2\,2|2\,2\rangle$, $\langle 1\sigma^2|H|2\sigma^2\rangle = \langle 1\,1|2\,2\rangle$

2. $\langle 1\sigma^2|H|1\sigma^2\rangle = (-2.6158)2 + 0.9596 = -4.2720$
$\langle 2\sigma^2|H|2\sigma^2\rangle = (-1.3154)2 + 0.6159 = -2.0149$
$\langle 1\sigma^2|H|2\sigma^2\rangle = 0.1261$

3. $E_- = -4.2790$, $E_+ = -2.0079$, $\mathbf{C}_- = (0.9984, -0.0556)$, $\mathbf{C}_+ = (0.0556, 0.9984)$

4.
$$\tfrac{1}{2}[|(a^{1/2}\phi_1 + b^{1/2}\phi_2)\alpha(a^{1/2}\phi_1 - b^{1/2}\phi_2)\beta| + |(a^{1/2}\phi_1 - b^{1/2}\phi_2)\alpha(a^{1/2}\phi_1 + b^{1/2}\phi_2)\beta|]$$
$$= \frac{1}{2\sqrt{2}}[(a^{1/2}\phi_1 + b^{1/2}\phi_2)(a^{1/2}\phi_1 - b^{1/2}\phi_2) + (a^{1/2}\phi_1 - b^{1/2}\phi_2)(a^{1/2}\phi_1 + b^{1/2}\phi_2)](\alpha\beta - \beta\alpha)$$
$$= (a\phi_1\phi_1 - b\phi_2\phi_2)(\alpha\beta - \beta\alpha)/\sqrt{2}$$
$$= a|\phi_1\alpha\phi_1\beta| - b|\phi_2\alpha\phi_2\beta|$$

2.4

1. The singlet function with $1\sigma 2\sigma$ occupancy is
$$\frac{1}{\sqrt{2}}[|1\,\alpha 2\,\beta| - |1\,\beta 2\,\alpha|]$$

2.
$$\langle 1\sigma 2\sigma|H|1\sigma^2\rangle = \frac{1}{\sqrt{2}}[2h_{12} + 2\langle 11|21\rangle] = \sqrt{2}[0.1954 - 0.1954] = 0$$
$$\langle 1\sigma 2\sigma|H|2\sigma^2\rangle = \frac{1}{\sqrt{2}}[2h_{12} + 2\langle 22|21\rangle] = \sqrt{2}[0.1954 - 0.0045] = 0.2699$$
$$\langle 1\sigma 2\sigma|H|1\sigma 2\sigma\rangle = h_{11} + h_{22} + \langle 12|12\rangle + \langle 12|21\rangle = -2.6158 - 1.3154 + 0.6063 + 0.1261 = -3.1988$$

The other matrix elements are derived in Problem 2.3.

3. The 3 × 3 CI matrix

$$\begin{pmatrix} -4.2720 & 0.1261 & 0.000 \\ & -2.0149 & 0.2699 \\ & & -3.1988 \end{pmatrix}$$

then has as its eigenvalues $E_1 = -4.2792, E_2 = -3.2567$, and $E_3 = -1.9497$. The corresponding eigenvectors have, as coefficients of $|1\sigma^2|$, $|2\sigma^2|$, and $|1\sigma 2\sigma|$, (0.9982, −0.0573, 0.0143), (−0.0261, −0.2098, 0.9772), and (0.0530, 0.9761, −0.2109), respectively. The ground-state total energy, including nuclear repulsion, is $-4.2792 + (2/1.4) = -2.8506$.

2.5 First iteration:

$$E^0 - E = -(\mathbf{C}^0)^T \mathsf{V} \mathbf{C}^0 = 0$$

and so we use $E = -4.2720$ in the first iteration to calculate $\mathbf{C}^{(1)}$:

$$\mathbf{C}^{(1)} = (E - \mathsf{H}^0)^{-1} \mathsf{V} \mathbf{C}^0, \qquad (E1 - \mathsf{H}^0)^{-1} = \begin{pmatrix} 0 & 0 & 0 \\ 0 & -0.4430 & 0 \\ 0 & 0 & -0.9318 \end{pmatrix}$$

(Actually, the 1,1 element of this matrix is ill defined. However, this does not cause trouble here since $\mathsf{V}\mathbf{C}^0$ has zero as its first entry. In general, however, this trouble arises wherever E^0 is taken to a diagonal element of H^0):

$$\mathsf{V}\mathbf{C}^0 = \begin{pmatrix} 0.0 \\ 0.1261 \\ 0.0 \end{pmatrix}$$

Therefore,

$$\mathbf{C}^{(1)} = \begin{pmatrix} 0.0 \\ -0.0559 \\ 0.0 \end{pmatrix}$$

A problem now arises: $\mathbf{C}^{(1)}$ does not obey intermediate normalization, which was assumed in deriving Eq. (2.110) for E. Hence we must damp the iteration process by averaging $\mathbf{C}^{(1)}$ and $\mathbf{C}^0$ to obtain a better $\mathbf{C}^{(1)}$ (which must then be intermediate normalized):

$$\mathbf{C}^{(1)} = \frac{1}{2}\begin{pmatrix} 1.0000 \\ 0.0 \\ 0.0 \end{pmatrix} + \frac{1}{2}\begin{pmatrix} 0.0 \\ -0.0559 \\ 0.0 \end{pmatrix} \rightarrow \begin{pmatrix} 1.0000 \\ -0.0559 \\ 0.0 \end{pmatrix}$$

One finds that by not damping, the successive $\mathbf{C}^{(n)}$ computed (using $E = -4.2790$) are wildly oscillating.

$$C^{(n)} = [(0.9989, 0, 0.0140), (0.0, -0.0573, -0.0035), (1.0000, 0.0004, 0.0139),$$
$$(1.000, 0.0797, 0.0001), (1.0000, 0.0386, 0.0138)]$$

However, once the damping is introduced, we can proceed to find $\mathbf{C}^{(n)}$ values in a stable manner.

$$E^0 - E = -(\mathbf{C}^0)^+ \mathsf{V}\mathbf{C} = -(0, 0.1261, 0)\begin{pmatrix} 1.0000 \\ -0.0559 \\ 0.0000 \end{pmatrix} = 0.0070$$

and so $E = E^0 - 0.0070 = -4.2790$. Now compute $\mathbf{C}^{(2)}$:

$$(E\mathsf{1} - \mathsf{H}^0)^{-1} = \begin{pmatrix} -142.9 & 0 & 0 \\ 0 & -0.4417 & 0 \\ 0 & 0 & -0.9258 \end{pmatrix}$$

and

$$\mathbf{C}^{(2)} = \begin{pmatrix} -142.9 & 0 & 0 \\ 0 & -0.4417 & 0 \\ 0 & 0 & -0.9258 \end{pmatrix}\begin{pmatrix} 0 & 0.1261 & 0 \\ 0.1261 & 0 & 0.2699 \\ 0 & 0.2699 & 0 \end{pmatrix}\begin{pmatrix} 1.0000 \\ -0.0559 \\ 0.0000 \end{pmatrix}$$

Then

$$\mathbf{C}^{(2)} = \begin{pmatrix} 1.0003 \\ -0.0557 \\ 0.0140 \end{pmatrix}$$

which upon intermediate normalization becomes

$$\mathbf{C}^{(2)} = \begin{pmatrix} 1.0000 \\ -0.0557 \\ 0.0140 \end{pmatrix}$$

For the third iteration,

$$E^0 - E = -(0, 0.1261, 0)\begin{pmatrix} 1.0000 \\ -0.0557 \\ 0.0140 \end{pmatrix} = 0.0070$$

Therefore, $E = -4.2790$ and so $(\mathrm{E}\mathsf{1} - \mathsf{H}^0)^{-1}$ is unchanged:

$$\mathbf{C}^{(3)} = (E\mathsf{1} - \mathsf{H}^0)^{-1}\begin{pmatrix} 0 & 0.1261 & 0 \\ 0.1261 & 0 & 0.2699 \\ 0 & 0.2699 & 0 \end{pmatrix}\begin{pmatrix} 1.0000 \\ -0.0557 \\ 0.0140 \end{pmatrix} = \begin{pmatrix} 1.0003 \\ -0.0574 \\ 0.0139 \end{pmatrix}$$

which when renormalized becomes

$$\begin{pmatrix} 1.0000 \\ -0.0574 \\ 0.0139 \end{pmatrix}$$

To compare these results with those of the CI Problem 2.4, we must normalize $\mathbf{C}^{(3)}$ so that $1 = (\mathbf{C}^{(3)})^{\mathrm{T}}(\mathbf{C}^{(3)})$. This procedure gives

$$\mathbf{C}^{(3)} = \begin{pmatrix} 0.9983 \\ -0.0573 \\ 0.0139 \end{pmatrix}$$

which is to be compared with the CI eigenvector

$$\mathbf{C}_{\mathrm{CI}} = \begin{pmatrix} 0.9982 \\ -0.0573 \\ 0.0143 \end{pmatrix}$$

The CI energy -4.2792 compares well with our third iterate $E = -4.2790$.

2.6

1. Let 1 and 2 denote the 1σ and 2σ orbitals, respectively:

	$\langle i| = \langle 0|$	$\langle i| = \langle 1|$
$\langle i|1_\alpha^+ 1_\alpha|0\rangle$	0.9970	0.0555
$\langle i|2_\alpha^+ 2_\alpha|0\rangle$	0.0031	−0.0555
$\langle i|1_\alpha^+ 1_\beta^+ 1_\beta 1_\alpha|0\rangle$	0.9970	0.0555
$\langle i|1_\alpha^+ 1_\beta^+ 2_\beta 2_\alpha|0\rangle$	−0.0555	−0.0031
$\langle i|2_\alpha^+ 2_\beta^+ 1_\beta 1_\alpha|0\rangle$	−0.0555	0.9970
$\langle i|2_\alpha^+ 2_\beta^+ 2_\beta 2_\alpha|0\rangle$	0.0031	−0.0555

Since the states $|0\rangle$, $|1\rangle$ have singlet symmetry, interchange of α and β spin gives the same matrix elements, e.g., $\langle i|1_\alpha^+ 1_\beta^+ 1_\beta 1_\alpha|0\rangle = \langle i|1_\beta^+ 1_\alpha^+ 1_\alpha 1_\beta|0\rangle$. Change of the sequence of the creation (or the annihilation) operators does, of course, change the sign of the matrix elements.

2. $V_{|1\rangle} = \langle 1|H|0\rangle = 0$, since $|1\rangle$ and $|0\rangle$ are determined from a CI calculation:

$$\begin{aligned} W_{12} &= \langle 0|[1_\alpha^+ 2_\alpha + 1_\beta^+ 2_\beta, H]|0\rangle = 2[h_{12}\langle 0|1_\alpha^+ 1_\alpha - 2_\alpha^+ 2_\alpha|0\rangle \\ &\quad + \langle 22|21\rangle\langle 0|2_\beta^+ 2_\alpha^+ 2_\beta 2_\alpha + 1_\alpha^+ 1_\beta^+ 2_\beta 2_\alpha|0\rangle \\ &\quad + \langle 11|21\rangle\langle 0|1_\alpha^+ 1_\beta^+ 1_\beta 1_\alpha + 1_\beta^+ 1_\alpha^+ 2_\beta 2_\alpha|0\rangle] \\ &= -0.0224 \end{aligned}$$

$$\begin{aligned}A_{21,21} &= \langle 0|[1_\alpha^+2_\alpha + 1_\beta^+2_\beta, H, 2_\alpha^+1_\alpha + 2_\beta^+1_\beta]|0\rangle\\ &= 2[(h_{22}-h_{11})\langle 0|1_\alpha^+1_\alpha - 2_\alpha^+2_\alpha|0\rangle - \langle 22|22\rangle\langle 0|2_\beta^+2_\alpha^+2_\alpha 2_\beta|0\rangle\\ &\quad - \langle 11|11\rangle\langle 0|1_\alpha^+1_\beta^+1_\beta 1_\alpha|0\rangle - 2\langle 11|22\rangle\langle 0|1_\beta^+1_\alpha^+2_\alpha 2_\beta|0\rangle\\ &\quad - \langle 21|12\rangle\langle 0|1_\beta^+1_\alpha^+1_\beta 1_\alpha + 2_\beta^+2_\alpha^+2_\beta 2_\alpha|0\rangle\\ &\quad + \langle 21|21\rangle\langle 0|1_\beta^+1_\alpha^+1_\alpha 1_\beta + 2_\alpha^+2_\beta^+2_\beta 2_\alpha|0\rangle]\\ &= 2.1606\end{aligned}$$

$$\begin{aligned}B_{21,21} &= \langle 0|[1_\alpha^+2_\alpha + 1_\beta^+2_\beta, H, 1_\alpha^+2_\alpha + 1_\beta^+2_\beta]|0\rangle\\ &= 2[(2\langle 22|11\rangle + 2\langle 21|21\rangle - \langle 11|11\rangle - \langle 22|22\rangle)\langle 0|1_\alpha^+1_\beta^+2_\beta 2_\alpha|0\rangle\\ &\quad - \langle 22|11\rangle\langle 0|2_\alpha^+2_\beta^+2_\beta 2_\alpha + 1_\alpha^+1_\beta^+1_\beta 1_\alpha|0\rangle]\\ &= -0.2400\end{aligned}$$

$$\begin{aligned}(\mathsf{A}_{21})_{|1\rangle,21} &= \langle 1|[H, 2_\alpha^+1_\alpha + 2_\beta^+1_\beta]|0\rangle = 2[h_{12}\langle 1|1_\alpha^+1_\alpha - 2_\alpha^+2_\alpha|0\rangle\\ &\quad - \langle 11|12\rangle\langle 1|1_\beta^+1_\alpha^+1_\beta 1_\alpha + 2_\alpha^+2_\beta^+1_\beta 1_\alpha|0\rangle\\ &\quad - \langle 22|12\rangle\langle 1|2_\beta^+2_\alpha^+1_\beta 1_\alpha + 2_\alpha^+2_\beta^+2_\beta 2_\alpha|0\rangle]\\ &= 0.4020\end{aligned}$$

$$\begin{aligned}(\mathsf{B}_{21})_{|1\rangle,21} &= \langle 1|[H, 1_\alpha^+2_\alpha + 1_\beta^+2_\beta]|0\rangle\\ &= 2[h_{12}\langle 1|2_\alpha^+2_\alpha - 1_\alpha^+1_\alpha|0\rangle - \langle 22|21\rangle\langle 1|2_\beta^+2_\alpha^+2_\beta 2_\alpha + 1_\alpha^+1_\beta^+2_\beta 2_\alpha|0\rangle\\ &\quad - \langle 11|21\rangle\langle 1|1_\alpha^+1_\beta^+1_\beta 1_\alpha + 1_\beta^+1_\alpha^+2_\beta 2_\alpha|0\rangle]\\ &= -0.0198\end{aligned}$$

$$(\mathsf{A}_{22})_{|1\rangle,|1\rangle} = -2.0079 + 4.2790 = 2.2711, \qquad (\mathsf{B}_{22})_{|1\rangle,|1\rangle} = 0$$

3. Since $\mathsf{A} - \mathsf{B}$ is a 2×2 matrix, we invert it easily and obtain the numerical value of κ_{21} and P_{10} through $\binom{\boldsymbol{\kappa}}{\mathbf{P}} = (\mathsf{A} - \mathsf{B})^{-1}\binom{\mathbf{W}}{\mathbf{V}}$. Hence the $\boldsymbol{\kappa}$ and P matrices are

$$\boldsymbol{\kappa} = \begin{pmatrix} 0.0 & 0.0097 \\ -0.0097 & 0.0 \end{pmatrix}, \qquad \mathsf{P} = \begin{pmatrix} 0.0 & -0.0018 \\ 0.0018 & 0.0 \end{pmatrix}$$

4. 'By applying $\exp(-\boldsymbol{\kappa})$ as shown in Problem 1.5 to the HF orbitals, we obtain

$$\phi_1 = 0.8919\ 1s_{\mathrm{He}} + 0.1701\ 1s_{\mathrm{H}}, \qquad \phi_2 = -0.8410\ 1s_{\mathrm{He}} + 1.2140\ 1s_{\mathrm{H}}$$

Likewise, transformation of the two CI eigenstates through $\exp(-\mathsf{P})$ gives rise to two new MC state vectors whose expansion coefficients are given by

$$\mathsf{C} = \begin{pmatrix} 0.9984 & 0.0574 \\ -0.0574 & 0.9984 \end{pmatrix}$$

5. Since we used the states of a CI calculation as the initial guess for the reference state in the one-step MCSCF procedure, the orbitals of the one- and two-step MCSCF approach become identical when the first iteration is carried out. The states obtained in the next iteration of the two-step MCSCF approach would, however, be determined from a new CI calculation (which requires transformed integrals) and would thus differ from the states obtained in the one-step MCSCF procedure.

6.

$E_{MCSCF} - E_{MCSCF}^{CONV}$	$(E_{MCSCF} - E_{MCSCF}^{CONV})^2$
2.1494×10^{-4}	4.6×10^{-8}
1×10^{-8}	1×10^{-16}

The error in the $(n + 1)$th iteration is the square of the error in the nth iteration.

7. The MCSCF and the full CI calculation have the same number and kind of variational parameters; hence the total energies obtained in the two calculations should become identical. The $|1\sigma 2\sigma\rangle$ configuration included in the CI wavefunction is treated in the two-configuration MCSCF function through the $2^+1\ \kappa_{21}$ orbital optimization parameter.

8. In an MCSCF calculation that uses $1\sigma^2$, $2\sigma^2$, and $1\sigma^1 2\sigma^1$, the $2\sigma^+1\sigma$ excitation operator becomes a redundant excitation operator and hence the orbital optimization step need not be included. The three-configuration MCSCF calculation thus becomes identical to the three-configuration CI calculation.

9. Again, two configurations plus one degree of orbital optimization freedom span all of the configuration space needed to generate the full CI wavefunction. Hence the converged MCSCF energy would equal the full CI energy here.

2.7

1. In the one- and two-electron integrals appearing in the second-quantized form of H, we neglect all integrals involving orbitals on different HeH^+ ions. Hence

$$H = \sum_{A=1}^{n} \left(\sum_{\substack{i,j=1\sigma,2\sigma \\ \in A}} h_{ij} i^+ j + \frac{1}{2} \sum_{\substack{ijkl=1\sigma,2\sigma \\ \in A}} \langle ij|kl\rangle i^+ j^+ lk \right) = \sum_{A=1}^{n} H_A$$

where A labels the n HeH^+ ions.

2. Let us denote the reference HF determinant by HF. Then doubly excited configurations involving excitation of the Ath HeH^+ ion can be represented as $(2_\alpha^+ 2_\beta^+ 1_\beta 1_\alpha)_A |HF\rangle \equiv |A\rangle$. Doubly excited configurations in which one orbital is excited on each of two HeH^+ ions will not give rise to

nonvanishing CI matrix elements since the one- and two-electron integrals that arise in evaluating such CI elements would vanish because of the large separation between the two ions.

The CI matrix elements arising from the functions $|\text{HF}\rangle$ and $\{|A\rangle\}$ are

$$\langle \text{HF}|H|\text{HF}\rangle = \sum_{A=1}^{n} \langle \text{HF}|H_A|\text{HF}\rangle = n\langle 1\sigma^2|H|1\sigma^2\rangle \equiv nE_{\text{HF}}$$

(This result follows since $|\text{HF}\rangle = \prod_{A=1}^{n}(1_\beta^+ 1_\alpha^+)_A|\text{vac}\rangle$);

$$\langle \text{HF}|H|A\rangle = \langle 1\sigma^2|H|2\sigma^2\rangle \qquad \text{(the same for all } A\text{)}$$

$$\langle A|H|A'\rangle = \delta_{AA'}[\langle 2\sigma^2|H|2\sigma^2\rangle + (n-1)E_{\text{HF}}] \qquad \text{(the same for all } A\text{)}$$

3. The components of the eigenvalue problem ($\mathsf{H}\mathbf{C} = E\mathbf{C}$) for the matrix shown in question 2 can be written as

$$nE_{\text{HF}}C_{\text{HF}} + \sum_{A=1}^{n} BC_A = EC_{\text{HF}}, \qquad BC_{\text{HF}} + CC_{A'} = EC_{A'}, \quad A' = 1, \ldots, n$$

Solving for $C_{A'}$ in terms of C_{HF} and substituting into the first equation gives

$$nE_{\text{HF}}C_{\text{HF}} + \sum_{A=1}^{n} B[E - C]^{-1}BC_{\text{HF}} = EC_{\text{HF}}$$

This equation will have a nontrivial solution for C_{HF} only if

$$nE_{\text{HF}} + B^2 n(E - C)^{-1} = E$$

This quadratic equation can be written as

$$(E - C)(nE_{\text{HF}} - C) + nB^2 = (E - C)^2$$

the solutions of which are

$$E - C = \tfrac{1}{2}\{nE_{\text{HF}} - C \pm [(nE_{\text{HF}} - C)^2 + 4nB^2]^{1/2}\}$$

Using the definitions of C and B, the ground-state energy becomes

$$E = nE_{\text{HF}} + \tfrac{1}{2}\{-(E_{\text{HF}}^* - E_{\text{HF}}) - [(E_{\text{HF}}^* - E_{\text{HF}})^2 + 4n\langle 11|22\rangle^2]^{1/2}\} + E_{\text{HF}}^* - E_{\text{HF}}$$

The correlation energy then becomes

$$E_{\text{corr}} = E - nE_{\text{HF}} = \tfrac{1}{2}\{E_{\text{HF}}^* - E_{\text{HF}} - [(E_{\text{HF}}^* - E_{\text{HF}})^2 + 4n\langle 11|22\rangle^2]^{1/2}\}$$

4. From Problem 2.1 we find $\langle 11|22\rangle = 0.1261$ and from Problem 2.3, $\langle 1\sigma^2|H|1\sigma^2\rangle = E_{\text{HF}} = -4.2720$ and $\langle 2\sigma^2|H|2\sigma^2\rangle = E_{\text{HF}}^* = -2.0149$. Therefore ($E_{\text{HF}}^* - E_{\text{HF}} = 2.2571$),

$$E_{\text{corr}} = \frac{2.2571}{2} - \left[\left(\frac{2.2571}{2}\right)^2 + n(0.1261)^2\right]^{1/2}$$

Evaluating E_{corr} for $n = 2, 4, 10, 100$, and 1000, and comparing it to n times the correlation energy of 1 HeH^+ ion, we find

n	E_{corr}	nE_{corr} $(n = 1)$	Error (%)
1	−0.0070	−0.0070	0.0
2	−0.0140	−0.0140	0.0*
4	−0.0278	−0.0280	0.7
10	−0.0684	−0.0700	2.3
100	−0.5637	−0.7000	19.5
1000	−3.0156	−7.0000	56.9

* If one were to carry more significant figures, this result would be 0.3%.

For large n, the analytical expression of question 3 clearly varies as $n^{1/2}$. Comparing our results for $n = 100$ and $n = 1000$, we find a ratio of $3.0156/0.5637 = 5.35$, which is not $(1000/100)^{1/2} = 3.16$. Thus $n = 100$ is not yet in the large-n range. The ratio for $n = 10$ and $n = 100$ is $0.5637/0.0684 = 8.24$, which is even further from $(100/10)^{1/2} = 3.16$. Hence one must go beyond $n = 100$ before this large-n behavior is realized.

5. Within our small basis the HeH^+ is undercorrelated because the 2σ orbital is much higher in energy than would be expected for the lowest excited σ orbital of HeH^+. Therefore, our correlation energy, which arises from the $1\sigma^2 \rightarrow 2\sigma^2$ excitation is smaller (because $\langle 11|22\rangle$ is smaller and $2\varepsilon_2 - 2\varepsilon_1$ is larger) than one would obtain if one were to use a better atomic orbital basis on HeH^+. As a result the $\langle 11|22\rangle$ appearing in the above expression for E_{corr} is "too small" and $(E^*_{HF} - E_{HF})$ is "too large." This leads to an underestimate of E_{corr}.

References

Banerjee, A., and Grein, F. (1976). *Int. J. Quantum Chem.* **10**, 123.
Bartlett, R. J., and Purvis, G. D. (1981). *Ann. N.Y. Acad. Sci.* **62**, 367.
Bender, C. F., and Davidson, E. R. (1967). *J. Chem. Phys.* **47**, 4972.
Brooks, B. R., and Schaefer, H. F. (1979). *J. Chem. Phys.* **70**, 5092.
Buenker, R. J., and Peyerimhoff, S. D. (1974). *Theor. Chim. Acta* **35**, 33.
Dalgaard, E., and Jørgensen, P. (1978). *J. Chem. Phys.* **69**, 3833.
Dalgaard, E. (1979). *Chem. Phys. Lett.* **65**, 559.
Davidson, E. R. (1975). *J. Comput. Phys.* **17**, 87.
Eyring, H., Walter, J., and Kimball, G. E. (1944). "Quantum Chemistry." Wiley, New York.
Kelly, H. P. (1964). *Phys. Rev. B* **136**, 896.
Löwdin, P. O. (1955). *Phys. Rev.* **97**, 1474.
Löwdin, P. O. (1968). *Int. J. Quantum Chem.* **2**, 867.

McWeeney, R., and Sutcliffe, B. T. (1976). "Methods of Molecular Quantum Mechanics." Academic Press, London.
Paldus, J., and Boyle, M. J. (1980). *Phys. Scripta* **21**, 295.
Pople, J. A., Seeger, R., and Krishnan, R. (1977). *Int. J. Quantum Chem.* **S11**, 149.
Roothaan, C. C. J. (1951). *Rev. Mod. Phys.* **23**, 69.
Roothaan, C. C. J. (1960). *Rev. Mod. Phys.* **32**, 179.
Schaefer, H. F., and Miller, W. H., Eds. (1977). "Modern Theoretical Chemistry," Vol. 3. Plenum, New York.
Shavitt, I. (1978). *Int. J. Quantum Chem.* **S12**, 5.
Yeager, D. L., Albertsen, P., and Jørgensen, P. (1980). *J. Chem. Phys.* **73**, 2811.
Yeager, D. L., and Jørgensen, P. (1979). *J. Chem. Phys.* **71**, 755.

Chapter 3 | Perturbation Theory

A. INTRODUCTION

As an alternative to the variational approaches described in Chapter 2, we may use perturbative approaches to solve the Schrödinger equation. Our purpose in this chapter is not to survey the many developments and applications that have been made of perturbation methods. Rather we attempt to cast the most familiar perturbation theory (Rayleigh-Schrödinger, RSPT) in the language of second quantization and then demonstrate how this tool can be used to compute state energies and wavefunctions. In Chapter 5 we extend this treatment to property average values and second-order response properties. We also illustrate some of the strengths and weaknesses of RSPT by comparing it with other perturbative and nonperturbative methods.

To begin, let us assume that the total electronic Hamiltonian H is decomposed into two pieces

$$H = H^0 + U \tag{3.1}$$

the former of which is assumed to be "larger" in a sense that will be clarified shortly. We also assume that we have available the complete set of eigenstates of H^0 (including the continuum, in principle)

$$H^0|k^0\rangle = E_k^0|k^0\rangle \tag{3.2}$$

This latter assumption often places substantial practical restrictions on the forms of H^0 that are possible. For example, choosing H^0 to be the N-electron Fock operator (or its second-quantized equivalent) would be quite reasonable because the $|k^0\rangle$ are then the usual Slater determinantal wavefunctions, which one is often actually able to obtain to reasonably high precision. On the other hand, choosing H^0 to include some factors (e.g., r_{ij}^{-1}) of the interelectronic distance is probably not practical because one cannot usually obtain the eigenstates $\{|j^0\rangle\}$ to high precision. Of course, the physical context and requisite accuracy relevant to each specific problem must ultimately dictate how one can most effectively split H into $H^0 + U$.

B. DERIVATION OF GENERAL ENERGY AND WAVEFUNCTION EXPRESSIONS

Given the above decomposition of H, we now express the Schrödinger equation

$$H|j\rangle = E_j|j\rangle \tag{3.3}$$

as (March *et al.*, 1967)

$$(\mathscr{E} - H^0)|j\rangle = (\mathscr{E} - E_j + U)|j\rangle \tag{3.4}$$

where the energy parameter $\mathscr{E}$, which has simply been added and subtracted, will be used shortly to define different kinds of perturbation theories. The normalization of the exact state $|j\rangle$ will now be chosen such that $|j\rangle$ has unit projection along its zeroth-order component $|j^0\rangle$

$$|j^0\rangle\langle j^0|j\rangle = |j^0\rangle \tag{3.5}$$

The total wavefunction $|j\rangle$ can now be written in terms of the projector Q,

$$Q \equiv 1 - |j^0\rangle\langle j^0|$$

as

$$|j\rangle = |j^0\rangle + Q|j\rangle \tag{3.6}$$

Making use of the fact that $H^0Q = QH^0$, which is easily seen to be valid from definition of Q, we can operate on Eq. (3.4) with the projector Q to obtain a closed expression for $Q|j\rangle$:

$$Q|j\rangle = (\mathscr{E} - H^0)^{-1}Q(\mathscr{E} - E_j + U)|j\rangle \tag{3.7}$$

The factor of Q, which when applied to Eq. (3.4) gave Eq. (3.7), is needed because the resolvent $(\mathscr{E} - H^0)^{-1}$ is singular at $\mathscr{E} = E_j^0$. The presence of Q guarantees that $(\mathscr{E} - H^0)^{-1}$ never operates on a state ($|j^0\rangle$) that would cause a singularity at this value of E_j^0. Clearly $(\mathscr{E} - H^0)^{-1}$ is singular at other values of $\mathscr{E}$ ($\mathscr{E} = E_k^0, k \neq j$) but we need not be concerned with these singularities as long as the parameter $\mathscr{E}$ is held in the neighborhood of E_j^0 and away from the other E_k^0. This would, of course, be difficult to achieve in systems such as metals, which possess many closely spaced (nearly degenerate) energy levels. Equation (3.7) is then inserted into Eq. (3.6) to yield the integral equation

$$|j\rangle = |j^0\rangle + (\mathscr{E} - H^0)^{-1}Q(\mathscr{E} - E_j + U)|j\rangle \tag{3.8}$$

If Eq. (3.8) is iterated, one obtains an order-by-order expansion of the wavefunction

$$|j\rangle = \sum_{n=0}^{\infty} [(\mathscr{E} - H^0)^{-1}Q(\mathscr{E} - E_j + U)]^n|j^0\rangle \tag{3.9}$$

Multiplying Eq. (3.3) on the left by $\langle j^0|$ we get

$$E_j = E_j^0 + \langle j^0|U|j\rangle \tag{3.10}$$

which then may be used to obtain a perturbative expansion of the energy in the perturbation U:

$$E_j - E_j^0 = \sum_{n=0}^{\infty} \langle j^0|U[(\mathscr{E} - H^0)^{-1}Q(\mathscr{E} - E_j + U)]^n|j^0\rangle \tag{3.11}$$

We now consider two especially relevant choices of the, in principle, arbitrary parameter $\mathscr{E}$. If $\mathscr{E}$ is taken to be equal to E_j, then the above perturbation series describe the Brillouin–Wigner approximations to $|j\rangle$ and E_j. The choice $\mathscr{E} = E_j^0$ yields the Rayleigh–Schrödinger perturbation series.

To express any of the above perturbation expansions in terms of creation and annihilation operators, we simply write $|j^0\rangle$, H^0, and U in the second-quantized manner. The zeroth-order Hamiltonian is virtually always taken to be a one-electron operator

$$H^0 = \sum_{k,l} \langle \phi_k|H^0|\phi_l\rangle k^+ l \tag{3.12}$$

involving a one-electron potential V,

$$H^0 = -\tfrac{1}{2}\nabla^2 - \sum_a Z_a|\mathbf{r} - \mathbf{R}_a|^{-1} + V \tag{3.13}$$

With this choice, the perturbation U becomes

$$U = \frac{1}{4}\sum_{ijkl} \langle ij||kl\rangle i^+ j^+ lk - \sum_{ij} \langle i|V|j\rangle i^+ j \tag{3.14}$$

If the spin-orbitals $\{\phi_k\}$ are chosen to diagonalize H^0,

$$H^0 = \sum_k \langle \phi_k|H^0|\phi_k\rangle k^+ k \equiv \sum_k \varepsilon_k k^+ k \tag{3.15}$$

then the zeroth-order wavefunctions $\{|j^0\rangle\}$ are simply N-electron Slater determinants (or their second-quantized equivalent) involving these same spin orbitals $\{\phi_k\}$. This choice of H^0 is especially convenient because the resolvent $(\mathscr{E} - H^0)^{-1}$ becomes diagonal (and hence easy to treat) within this representation.

To gain some experience in the evaluation of perturbation contributions to $|j\rangle$ and E_j and to motivate an analysis of a fundamental weakness of the Buillouin–Wigner perturbation theory (BWPT), let us now consider a few examples. First, we evaluate the first-order correction to the energy that arises from the $n = 0$ term in Eq. (3.11):

$$E_j^{(1)} = \langle j^0|U|j^0\rangle = \frac{1}{4}\sum_{ijkl}\langle j^0|i^+j^+lk|j^0\rangle\langle ij||kl\rangle - \sum_{kl}\langle j^0|k^+l|j^0\rangle\langle k|V|l\rangle \tag{3.16}$$

which by straightforward application of Slater–Condon-like rules to compute the above density matrices, yields

$$E_j^{(1)} = \frac{1}{2}\sum_{\mu,\nu\in j_0}\langle\mu\nu||\mu\nu\rangle - \sum_{\mu\in j_0}\langle\mu|V|\mu\rangle. \tag{3.17}$$

Note that because $E_j^{(1)}$ contains no reference to the parameter $\mathscr{E}$, RSPT and BWPT have identical first-order energies.

C. SIZE CONSISTENCY PROBLEM IN THE ENERGY

Next, we consider the second-order energy and, in particular, we examine $E_j^{(2)}$ for a system consisting of two noninteracting subsystems (a and b). For this case, the Hamiltonian H separates into

$$H = H_a^0 + U_a + H_b^0 + U_b \tag{3.18}$$

and the zeroth-order states become (antisymmetric) product states (recall that $|j^0\rangle$ labels our specific state of interest whereas $|k^0\rangle$ labels the other states):

$$|j^0\rangle = |j_a^0 j_b^0\rangle, \{|k^0\rangle\} = \{|k_a^0 j_b^0\rangle, |j_a^0 k_b^0\rangle, |k_a^0 k_b^0\rangle\} \tag{3.19}$$

The second-order energy expression from Eq. (3.11) reduces, using $Q|j^0\rangle = 0$, to

$$E_j^{(2)} = \langle j^0|U(\mathscr{E} - H^0)^{-1}QU|j^0\rangle \tag{3.20}$$

Expressing Q in the conventional sum-over-states manner gives

$$Q \equiv 1 - |j^0\rangle\langle j^0| = \sum_{k^0\neq j^0}|k^0\rangle\langle k^0| \tag{3.21}$$

and Eq. (3.20) becomes

$$E_j^{(2)} = \sum_{k^0 \neq j^0} \langle j^0|U|k^0\rangle\langle k^0|U|j^0\rangle/(\mathscr{E} - E_k^0) \tag{3.22}$$

Now, introducing the fact that we are dealing with two noninteracting subsystems [through Eqs. (3.18) and (3.19)] and breaking the sum over k^0 into three sums corresponding to the partitioning of $|k^0\rangle$ given in Eq. (3.19), we obtain

$$\begin{aligned} E_j^{(2)} = &\sum_{k_a^0} \frac{|\langle j_a^0 j_b^0|U_a + U_b|k_a^0 j_b^0\rangle|^2}{\mathscr{E} - E_{j_b}^0 - E_{k_a}^0} + \sum_{k_b^0} \frac{|\langle j_a^0 j_b^0|U_a + U_b|j_a^0 k_b^0\rangle|^2}{\mathscr{E} - E_{j_a}^0 - E_{k_b}^0} \\ &+ \sum_{k_a^0, k_b^0} \frac{|\langle j_a^0 j_b^0|U_a + U_b|k_a^0 k_b^0\rangle|^2}{\mathscr{E} - E_{k_a}^0 - E_{k_b}^0} \end{aligned} \tag{3.23}$$

Using the orthonormality of the spin-orbitals and the fact that the systems are noninteracting permits the simplification of the above matrix elements. For example,

$$\langle j_a^0 j_b^0|U_a + U_b|k_a^0 j_b^0\rangle = \langle j_a^0|U_a|k_a^0\rangle, \qquad \langle j_a^0 j_b^0|U_a + U_b|k_a^0 k_b^0\rangle = 0 \tag{3.24}$$

Thus, we finally obtain

$$E_j^{(2)} = \sum_{k_a^0} \frac{|\langle j_a^0|U_a|k_a^0\rangle|^2}{\mathscr{E} - E_{j_b}^0 - E_{k_a}^0} + \sum_{k_b^0} \frac{|\langle j_b^0|U_b|k_b^0\rangle|^2}{\mathscr{E} - E_{j_a}^0 - E_{k_b}^0} \tag{3.25}$$

Notice that this total second-order energy is not, in general, a sum of the second-order energies of the two separated species because of the appearance of the E_{jb}^0 and E_{ja}^0 terms in the denominators. We therefore say that the general (i.e., with arbitrary $\mathscr{E}$) perturbation theory energy is not size consistent (Pople *et al.*, 1977). However, if $\mathscr{E}$ is chosen, as in RSPT, equal to $\mathscr{E} = E_j^0 = E_{ja}^0 + E_{jb}^0$, then we indeed obtain a perfectly size-consistent result:

$$E_j^{(2)} = E_{ja}^{(2)} + E_{jb}^{(2)} \tag{3.26}$$

It is clear that the first-order energy expression $\langle j^0|U|j^0\rangle$ is also size consistent. Because this size consistency property is important, especially if we are interested in using perturbation methods to study molecular fragmentation, the use of RSPT must be favored over BWPT ($\mathscr{E} = E_j$) or any other perturbation theory derived from alternative choices of $\mathscr{E}$.

Even if we now decided to use only RSPT to compute $|j\rangle$ and E_j, another potential difficulty arises when we consider the third- (and higher-) order energies

$$\begin{aligned} E_j^{(3)} = &\langle j^0|UQ(E_j^0 - H^0)^{-1}UQ(E_j^0 - H^0)^{-1}U|j^0\rangle \\ &- E_j^{(1)}\langle j^0|UQ(E_j^0 - H^0)^{-2}U|j^0\rangle \end{aligned} \tag{3.27}$$

Because $E_j^{(1)}$ is an extensive property (i.e., size consistent), we should expect possible size inconsistency from the last term in $E_j^{(3)}$ if the (second-order) factor in front of $E_j^{(1)}$ is also extensive; $E_j^{(3)}$ would then contain terms that are proportional to the square of the system's size (for identical noninteracting subsystems). Let us now look at this situation more closely. By introducing the spectral representation of the projector Q given in Eq. (3.21), $E_j^{(3)}$ can be written as

$$E_j^{(3)} = \sum_{k^0, l^0} \frac{\langle j^0|U|k^0\rangle\langle k^0|U|l^0\rangle\langle l^0|U|j^0\rangle}{(E_j^0 - E_k^0)(E_j^0 - E_l^0)} - \langle j^0|U|j^0\rangle \sum_{k^0} \frac{\langle j^0|U|k^0\rangle\langle k^0|U|j^0\rangle}{(E_j^0 - E_k^0)^2} \tag{3.28}$$

It is now important to demonstrate that the $k^0 = l^0$ terms appearing in the first sum above exactly cancel the size-inconsistent terms in the second factor. This cancellation can be brought about by combining these terms as

$$\Delta = \sum_{k^0} \frac{|\langle j^0|U|k^0\rangle|^2}{(E_j^0 - E_k^0)^2} [\langle k^0|U|k^0\rangle - \langle j^0|U|j^0\rangle] \tag{3.29}$$

Now, if we consider Δ for the special case of two noninteracting subsystems (analogous to what was done above for $E_j^{(2)}$), we see that the terms $\langle k^0|U|k^0\rangle - \langle j^0|U|j^0\rangle$ decompose, under partitioning of k^0 [as in Eq. (3.19)], into

$$\langle j_a^0 k_b^0|U_a + U_b|j_a^0 k_b^0\rangle - \langle j_a^0 j_b^0|U_b|j_a^0 j_b^0\rangle = \langle k_b^0|U_b|k_b^0\rangle - \langle j_b^0|U_b|j_b^0\rangle \tag{3.30}$$

(notice that reference to system a has disappeared here) and

$$\langle k_a^0 j_b^0|U_a + U_b|k_a^0 j_b^0\rangle - \langle j_a^0 j_b^0|U_a|j_a^0 j_b^0\rangle = \langle k_a^0|U_a|k_a^0\rangle - \langle j_a^0|U_a|j_a^0\rangle. \tag{3.31}$$

(reference to system b has disappeared here).

The states $\{|k_a^0 k_b^0\rangle\}$ give no contributions because the first factor in Eq. (3.29) $\langle j^0|U|k^0\rangle$ becomes identically zero for these states. This then permits Δ to be written as a sum of terms referring totally to system a:

$$\Delta_a = \sum_{k_a^0} \frac{|\langle j_a^0|U_a|k_a^0\rangle|^2}{(E_{j_a}^0 - E_{k_a}^0)^2} [\langle k_a^0|U_a|k_a^0\rangle - \langle j_a^0|U_a|j_a^0\rangle] \tag{3.32}$$

and an analogous expression for Δ_b. Hence Δ is size consistent even though each of the two terms arising in it [see Eq. (3.29)] are not. When $k^0 \neq l^0$ the first term in Eq. (3.28) may also be shown to be size consistent and $E_j^{(3)}$ therefore is size consistent. RSPT energies are in general size consistent even though substantial regrouping of terms as in $E_j^{(3)}$ may be necessary before it can be realized.

If one wishes to use RSPT to perform ab initio quantum-chemical calculations that yield size-consistent energies, then care must be taken in computing the factors that contribute to any given $E_j^{(n)}$. For example, if $E_j^{(3)}$ were calculated as in Eq. (3.28), limitations of numerical precision might not give rise to the exact cancellation of size-inconsistent terms, which we know should occur. This would certainly be the case for an extended system (for which the size-inconsistent terms would dominate). In addition, it is unpleasant to have a formalism in which such improper terms arise in the first place. It is therefore natural to attempt to develop approaches to implementing RSPT in which the size-inconsistent factors are never even computed. Such an approach has been developed and is commonly referred to as many-body perturbation theory (MBPT). The method of implementing MBPT is discussed once we have completed the present treatment of RSPT.

D. MØLLER–PLESSET PERTURBATION THEORY FOR ENERGY

A very common choice of the potential V used to define H^0 as in Eq. (3.13) is the HF potential

$$V = \sum_{k,l} \sum_{\mu \in j^0} \langle k\mu || l\mu \rangle k^+ l \tag{3.33}$$

where the sum over μ runs over those spin-orbitals that are occupied in the specific zeroth-order state (the Slater determinant $|j^0\rangle$) whose perturbation we are examining. The unperturbed Hamiltonian H^0 is then given in terms of the HF orbital energies as

$$H^0 = \sum_k \varepsilon_k k^+ k \tag{3.34}$$

With the above choice of H^0 now made, the perturbation U becomes

$$U = \frac{1}{4} \sum_{ijkl} \langle ij || kl \rangle i^+ j^+ lk - \sum_{k,l} \sum_{\mu \in jo} \langle k\mu || l\mu \rangle k^+ l \tag{3.35}$$

These choices of H^0 and U, when used in RSPT, give rise to what is commonly called (Pople *et al.*, 1977) Møller–Plesset perturbation theory (MPPT) and the expression for $E_j^{(1)}$ reduces to the familiar form

$$E_j^{(1)} = -\frac{1}{2} \sum_{\mu,\nu \in jo} \langle \mu\nu || \mu\nu \rangle \tag{3.36}$$

The RSPT expression for $E_j^{(2)}$ can also be expressed in terms of orbital energies and two-electron integrals. The kets $|k^0\rangle$ appearing in Eq. (3.22) for

$E_j^{(2)}$ refer to kets that are singly, doubly, etc. excited relative to the reference ket $|j^0\rangle$. Because of the BT, $\langle j^0|U|k^0\rangle$ vanishes when $|k^0\rangle$ is a singly excited ket. Kets $|k^0\rangle$ that are triply excited and higher also give no contribution to $\langle j^0|U|k^0\rangle$. Hence, $E_j^{(2)}$ as given in Eq. (3.22) would contain only contributions from the doubly excited kets,

$$\langle j^0|Ur^+s^+\beta\alpha|j^0\rangle = \langle\alpha\beta||rs\rangle \tag{3.37}$$

and therefore

$$E_j^{(2)} = -\sum_{\substack{\alpha<\beta\\ r<s}} \frac{|\langle\alpha\beta||rs\rangle|^2}{\varepsilon_r + \varepsilon_s - \varepsilon_\alpha - \varepsilon_\beta} \tag{3.38}$$

To obtain more insight into the structure of the size-consistent and -inconsistent terms, we derive an explicit expression for the quantity Δ appearing in Eq. (3.29) within MPPT. Realizing that only the doubly excited determinants $|k^0\rangle$ contribute in Eq. (3.29) because of the BT, we obtain

$$\Delta = \sum_{\substack{m<n\\ \alpha<\beta}} \frac{|\langle\alpha\beta||mn\rangle|^2}{(\varepsilon_\alpha + \varepsilon_\beta - \varepsilon_m - \varepsilon_n)^2}\left[\langle{}^{mn}_{\alpha\beta}|U|{}^{mn}_{\alpha\beta}\rangle - \langle j^0|U|j^0\rangle\right] \tag{3.39}$$

where

$$|{}^{mn}_{\alpha\beta}\rangle \equiv m^+n^+\beta\alpha|j^0\rangle \tag{3.40}$$

The expectation value difference contained in the square brackets can be expressed in terms of elementary two-electron integrals and, since $U = W - V$, HF potential matrix elements as

$$\sum_{\substack{\gamma\in j^0\\ \gamma\neq\alpha,\beta}} [\langle m\gamma||m\gamma\rangle + \langle n\gamma||n\gamma\rangle - \langle\alpha\gamma||\alpha\gamma\rangle - \langle\beta\gamma||\beta\gamma\rangle] + \langle mn||mn\rangle - \langle\alpha\beta||\alpha\beta\rangle$$
$$- \langle m|V|m\rangle - \langle n|V|n\rangle + \langle\alpha|V|\alpha\rangle + \langle\beta|V|\beta\rangle \tag{3.41}$$

By then introducing the explicit form of V given in Eq. (3.33) this expression can be reduced to

$$\langle mn||mn\rangle + \langle\alpha\beta||\alpha\beta\rangle - (1 + P_{mn})(1 + P_{\alpha\beta})\langle m\alpha||n\beta\rangle \tag{3.42}$$

where P_{mn} means interchanging the indices m and n. Upon inserting this into Eq. (3.39) one obtains

$$\Delta = \sum_{\substack{m<n\\ \alpha<\beta}} \frac{|\langle\alpha\beta||mn\rangle|^2}{(\varepsilon_\alpha + \varepsilon_\beta - \varepsilon_m - \varepsilon_n)^2}[\langle mn||mn\rangle + \langle\alpha\beta||\alpha\beta\rangle$$
$$- (1 + P_{mn})(1 + P_{\alpha\beta})\langle m\alpha||n\beta\rangle] \tag{3.43}$$

The expression for Δ given in Eq. (3.39) may be written as a product of two independent disjoint sums since part of the first term $\langle {}^{mn}_{\alpha\beta}|U|^{mn}_{\alpha\beta}\rangle$ and the whole second term $\langle j^0|U|j^0\rangle$ in the square bracket consists of a sum that is independent of the sum $m < n$ and $\alpha < \beta$. Equation (3.39) thus contains a product of two disjoint terms. When the terms in the square brackets are collected together in a different manner, Eq. (3.39) reduces to Eq. (3.43), which cannot be divided up into disjoint sums. The expression for Δ given in Eq. (3.43) is thus linked. It is the linked nature of the size-consistent terms that is used in MBPT to assure that size-consistent terms alone will appear in the many-body perturbation expressions for the electronic energy and other state properties.

E. THE PERTURBED WAVEFUNCTION

Having now carried out some detailed analysis of the RSPT expression for E_j, let us turn to the perturbative corrections to the wavefunction $|j\rangle$. The first-order RSPT wavefunction is, according to Eq. (3.9),

$$|j^{(1)}\rangle = (E_j^0 - H^0)^{-1}Q(E_j^0 - E_j + U)|j^0\rangle \tag{3.44}$$

which, because $Q|j^0\rangle = 0$, reduces to

$$|j^{(1)}\rangle = (E_j^0 - H^0)^{-1}QU|j^0\rangle \tag{3.45}$$

By inserting the spectral representation of Q [Eq. (3.21)], we obtain

$$|j^{(1)}\rangle = \sum_{k^0 \neq j^0} \frac{\langle k^0|U|j^0\rangle}{E_j^0 - E_k^0}|k^0\rangle \tag{3.46}$$

As we did above for the energy, it is instructive to analyze $|j^{(1)}\rangle$ when it pertains to two noninteracting subsystems (a and b). For this special case, the sum in Eq. (3.46) separates into terms pertaining to each of the isolated systems:

$$|j^{(1)}\rangle = -\sum_{k_a^0} \frac{\langle k_a^0|U_a|j_a^0\rangle}{E_{k_a}^0 - E_{j_a}^0}|k_a^0 j_b^0\rangle - \sum_{k_b^0} \frac{\langle k_b^0|U_b|j_b^0\rangle}{E_{k_b}^0 - E_{j_b}^0}|j_a^0 k_b^0\rangle \tag{3.47}$$

Thus, we see that, through first order, the wavefunction $|j\rangle$ contains only terms of the form $|j_a^0 j_b^0\rangle$, $|j_a^0 k_b^0\rangle$, and $|k_a^0 j_b^0\rangle$; terms such as $|k_a^0 k_b^0\rangle$ are not present. One might have expected that, for two noninteracting subsystems, the total wavefunction should be a (antisymmetric) product of the wavefunctions for each subsystem and that terms like $|k_a^0 k_b^0\rangle$ would, thus, be present. The RSPT first-order wavefunction certainly does not possess this product-separable property.

It is natural to ask how the RSPT can have the physically consistent property that its energy is extensive whereas its wavefunction does not reduce to a product form for noninteracting systems. The answer has to do with the manner in which the total energy E_j is computed in perturbation theory:

$$E_j = \langle j^0|H|j\rangle \tag{3.48}$$

In contrast, the total energy is obtained, in variational approaches, by evaluating the expectation value

$$E_j = \langle j|H|j\rangle/\langle j|j\rangle \tag{3.49}$$

Because of the structure of the matrix element appearing in Eq. (3.48) it is not possible for terms such as $|k_a^0 k_b^0\rangle$ to contribute directly to the RSPT expressions for E_j even though these factors are certainly contained in the exact wavefunction $|j\rangle$ (they will occur as higher order RSPT wavefunction corrections). That is,

$$\langle j_a^0 j_b^0|H|k_a^0 k_b^0\rangle \equiv \langle j_a^0 j_b^0|H_a^0 + H_b^0 + U_a + U_b|k_a^0 k_b^0\rangle = 0 \tag{3.50}$$

In a sense then, the first-order RSPT wavefunction contains faults (absence of $|k_a^0 k_b^0\rangle$) that do not adversely affect its ability to yield, through

$$E_j^{(2)} = \langle j^0|U|j^{(1)}\rangle \tag{3.51}$$

a size-consistent second-order energy. The same form of the wavefunction ($|j\rangle \cong |j^0\rangle + |j^{(1)}\rangle$) when used in the expectation value [Eq. (3.49)] would not yield a size-consistent result; one would have to add on the $|k_a^0 k_b^0\rangle$ terms to generate size consistency in the expectation value. This implies that in a CI or MCSCF calculation one must use these disjoint excitations $|k_a^0 k_b^0\rangle$ (e.g., $|\sigma_u^2 \sigma_u^2\rangle$ for two H_2 molecules) in order to guarantee that the expectation value formula for the energy is size consistent. This necessity that one include, in a CI study of a composite system, excitation levels (in Slater determinants) that are higher than those included for the individual constituent fragments is a problem of the CI and MCSCF methods.

F. MØLLER–PLESSET WAVEFUNCTION

If the unperturbed Hamiltonian H^0 is taken to be the HF Hamiltonian, the first-order MPPT wavefunction [Eq. (3.46)] can be expressed in terms of the doubly excited kets $r^+ s^+ \beta\alpha|j^0\rangle$ (again BT makes the contributions due to singly excited kets vanish):

$$|j^{(1)}\rangle = \sum_{\substack{\alpha<\beta \\ r<s}} K_{\alpha\beta}^{rs} r^+ s^+ \beta\alpha |j^0\rangle \tag{3.52}$$

where

$$K_{\alpha\beta}^{rs} = -\frac{\langle rs||\alpha\beta\rangle}{\varepsilon_r + \varepsilon_s - \varepsilon_\alpha - \varepsilon_\beta} \tag{3.53}$$

The explicit expression for the second-order wavefunction for this (HF) H^0 case contains singly, doubly, triply, and quadruply excited kets

$$\begin{aligned} |j^{(2)}\rangle = &\sum_{r,\alpha} K_\alpha^r r^+ \alpha |j^0\rangle + \sum_{\substack{\alpha<\beta \\ r<s}} K_{\alpha\beta}^{rs} r^+ s^+ \beta\alpha |j^0\rangle \\ &+ \sum_{\substack{\alpha<\beta<\gamma \\ r<s<t}} K_{\alpha\beta\gamma}^{rst} r^+ s^+ t^+ \gamma\beta\alpha |j^0\rangle + \sum_{\substack{\alpha<\beta<\gamma<\delta \\ r<s<t<u}} K_{\alpha\beta\gamma\delta}^{rstu} r^+ s^+ t^+ u^+ \delta\gamma\beta\alpha |j^0\rangle \end{aligned} \tag{3.54}$$

where, for example,

$$K_\alpha^r = \frac{1}{2}\left(\sum_{\substack{mn \\ \gamma}} \frac{\langle r\gamma||mn\rangle\langle mn||\gamma\alpha\rangle}{(\varepsilon_\alpha - \varepsilon_r)(\varepsilon_\gamma + \varepsilon_\alpha - \varepsilon_n - \varepsilon_m)} + \sum_{\substack{m \\ \gamma\beta}} \frac{\langle \beta\gamma||\alpha m\rangle\langle mr||\gamma\beta\rangle}{(\varepsilon_\alpha - \varepsilon_r)(\varepsilon_\gamma + \varepsilon_\beta - \varepsilon_r - \varepsilon_m)} \right) \tag{3.55}$$

G. MANY-BODY PERTURBATION THEORY

Having now completed our treatment of RSPT, let us return to the problem of finding a mechanism for explicitly computing, in any given order, only those terms in $E_j^{(n)}$ or $|j^{(n)}\rangle$ that are size consistent. Recall that RSPT, as normally expressed, contains size-inconsistent terms that cancel when grouped together properly, but that nevertheless appear in the formal RSPT expression. Recall also that the size-inconsistent terms could be characterized by a factorization into products of two or more terms that did not share common summation indices. In MBPT, the formal cancellation of size-inconsistent terms in RSPT is carried out explicitly (Brueckner, 1955a,b; Bartlett and Silver, 1975; Kelly, 1969; Löwdin, 1968; Brandow, 1977). Formally, we may thus write $E_j^{(3)}$ as

$$\begin{aligned} E_j^{(3)}(\text{RSPT}) &= \langle j^0|UQ(E_j^0 - H^0)^{-1}U(E_j^0 - H^0)^{-1}QU|j^0\rangle \\ &\quad - E_j^{(1)}\langle j^0|UQ(E_j^0 - H^0)^{-2}U|j^0\rangle \\ &\equiv \langle j^0|U(E_j^0 - H^0)^{-1}U(E_j^0 - H^0)^{-1}U|j^0\rangle_L = E_j^{(3)}(\text{MBPT}) \end{aligned} \tag{3.56}$$

or $E_j^{(n)}$ in general, using Eq. (3.11), as

$$E_j^{(n)} = \langle j^0 | U\{(E_j^0 - H^0)^{-1} QU\}^n | j^0 \rangle_L \tag{3.57}$$

The subscript L indicates that only the size-consistent terms in that expression are included. The term involving $E_j^{(1)}$ in Eq. (3.56) give rise to purely size-inconsistent terms. In MBPT, the size-consistent terms are said to be linked, while the size-inconsistent terms are referred to as unlinked. If all of the unlinked terms were trivial to identify, such as the second term in $E_j^{(3)}$(RSPT), one could merely exclude them. However, as we demonstrated earlier, there are also unlinked contributions in the first term in $E_j^{(3)}$(RSPT) that cancel those in the second term and that are not easily identified. It is the strength of MBPT that it allows us directly to identify all the linked terms of RSPT. The derivation that shows how to identify the size-consistent or linked terms is rather tedious and is described in detail in many textbooks (March *et al.*, 1967; Raimes, 1972; Linderberg and Öhrn, 1973). We do not carry out that derivation but simply familiarize the reader with the language of MBPT and report the results of the derivation. The implementation of the MBPT method for evaluating only the linked contributions to $E_j^{(n)}$ is commonly given in terms of a set of diagrams, the numerical values of which are the desired size-consistent components. We use the so-called Hugenholtz diagram rules to determine the number of diagrams, which enter in a given order (n) in the perturbation. In Table I, we report the rules for constructing these diagrams and in Fig. 3.1, the Hugenholtz energy diagrams that enter up through second order are displayed. The translation of the Hugenholtz diagrams into algebraic expressions is commonly performed by translating the Hugenholtz diagram into one of its equivalent Brandow diagrams (Brandow, 1977). The algebraic expression for the Hugenholtz diagram is then obtained by applying the rules given in Table II to the Brandow diagram.

Table I

Rules for Constructing All Hugenholtz Diagrams for a Given Order n

1. Represent each of the two-electron interactions (W) with a dot having two incoming and two outgoing lines (e.g., $\times$) and each of the one-electron terms ($-V$) with a solid line having one endpoint at which one line is entering and one leaving (e.g., $\succ\!\!\longrightarrow\!\!\bullet$) ($H = H_0 + W - V$).
2. To a given order n in the perturbation, write all possible ways (on a time axis) of drawing m ($m = 0, 1, \ldots, n$) dots and $n - m$ solid lines with one endpoint.
3. Connect the lines entering and leaving a dot and a solid line with one endpoint in all possible different ways such that the resulting diagrams are linked. A linked diagram cannot be pulled apart into two separate diagrams without cutting lines. An example of an unlinked diagram is given in Fig. 3.3A.

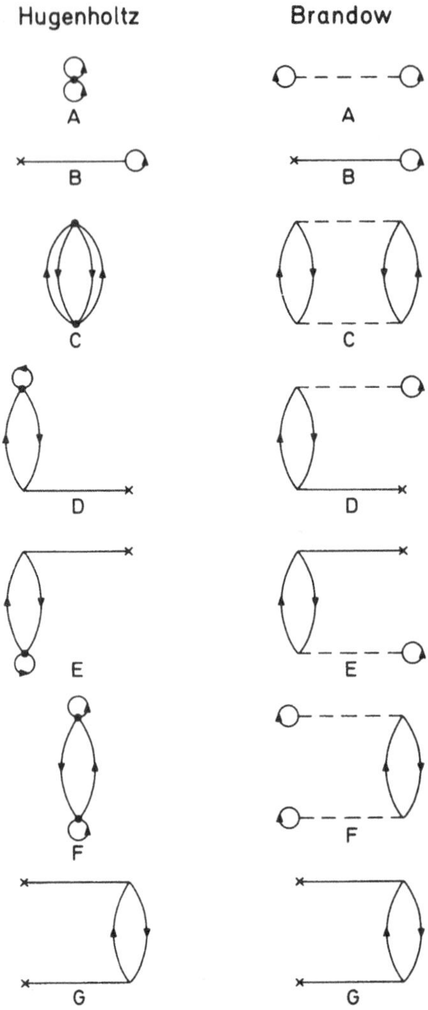

FIG. 3.1. First- and second-order Hugenholtz and Brandow energy diagrams.

A Hugenholtz diagram is translated into one of its equivalent Brandow diagrams when the dots (which represent the two-electron interaction) are extended into dashed lines, where one arrow is entering and one leaving at both end points of the dashed line, e.g., 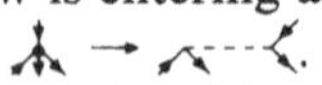. The Hugenholtz diagram in Fig. 3.2, may, for example, be translated into one of the eight Brandow diagrams given in Fig. 3.2. At first glance, these eight diagrams look very different, but when applying the rules in Table II, their algebraic expressions become identical. To illustrate this and to get some experience in applying the rules in Table II, we evaluate Brandow diagrams A and E

Table II

Rules for Evaluating Brandow Diagrams.

1. Label the diagram with general "hole" ($\alpha, \beta, \gamma, \ldots$) (down arrow) and "particle" ($m, n, p, \ldots$) (up arrow) indices. A line that starts and ends at the same interaction is labeled with a hole index. An example of the labeling is given in Fig. 3.2A,E.
2. The numerator of the diagram contains products of the one-electron integrals (if any) and the antisymmetrized two-electron integrals. The indices of the one- and two-electron integrals are assigned according to the rule

$$\langle \text{out}| - V|\text{in}\rangle$$

$$\langle \text{left-out, right-out}||\text{left-in, right-in}\rangle$$

Examples:

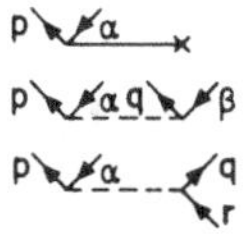

$\sim \quad \langle p| - V|\alpha\rangle$

$\sim \quad \langle pq||\alpha\beta\rangle$

$\sim \quad \langle pq||\alpha r\rangle$

3. The denominator corresponding to a given diagram is obtained by taking a factor equal to the sum of the hole orbital energies minus the sum of the particle orbital energies for each horizontal cut the eye draws between successive pairs of either dotted or solid lines. These $n - 1$ individual factors are then multiplied to form the denominator.
4. Multiply the diagram by $(\frac{1}{2})^n$, where n is the number of "equivalent pairs" of lines. Two lines form an equivalent pair if they both begin at the same interaction, both end at the same interaction, and both go in the same direction (e.g., Fig. 3.1C has two pairs, Fig. 3.3A has one pair, and Fig. 3.2A–H has no pairs).
5. Multiply each numerator by $(-1)^{l+h}$, where h is the number of hole lines in the diagram and l the number of closed loops. A closed loop is formed when one can trace from one endpoint of an interaction along the direction of an arrow and end up back at the same point without ever having to cross an interaction (dashed) line (Fig. 3.3A contains three loops, Fig. 3.2C contains two loops, and Fig. 3.2B has one loop).
6. Sum over all particle and hole states that occur in the diagram.

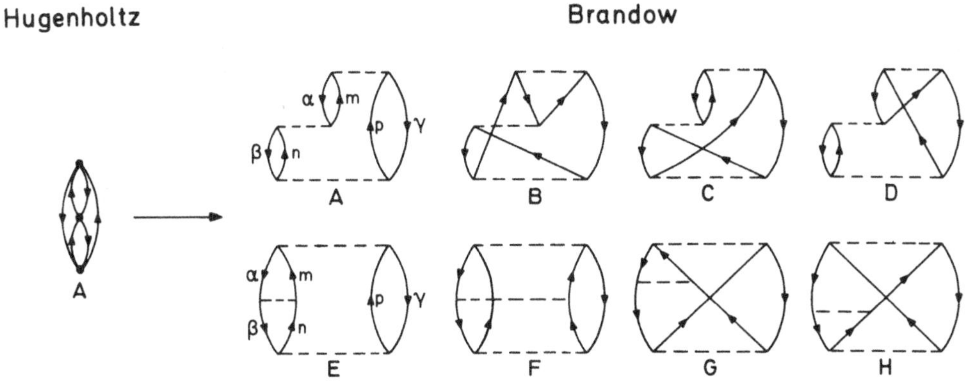

FIG. 3.2. Translation of a third-order Hugenholtz diagram into corresponding Brandow diagrams.

in Fig. 3.2. We obtain, using the labeling of particle and hole lines given in the figure

$$\mathrm{A} = \sum_{\substack{\alpha\beta\gamma \\ mnp}} (-1)^{3+3} \frac{\langle \alpha\gamma||mp\rangle\langle \beta m||n\alpha\rangle\langle np||\beta\gamma\rangle}{(\varepsilon_\alpha + \varepsilon_\gamma - \varepsilon_m - \varepsilon_p)(\varepsilon_\beta + \varepsilon_\gamma - \varepsilon_n - \varepsilon_p)} \tag{3.58}$$

Diagram A contains three hole lines and three closed loops. Diagram E may, in a similar way, be expressed as

$$\mathrm{E} = \sum_{\substack{\alpha\beta\gamma \\ mnp}} (-1)^{2+3} \frac{\langle \alpha\gamma||mp\rangle\langle \beta m||\alpha n\rangle\langle np||\beta\gamma\rangle}{(\varepsilon_\alpha + \varepsilon_\gamma - \varepsilon_m - \varepsilon_p)(\varepsilon_\beta + \varepsilon_\gamma - \varepsilon_n - \varepsilon_p)} \tag{3.59}$$

since E contains two closed loops. Interchanging α and n in the second electronic interaction in E gives a minus sign and A and E thus become identical.

The reason for including only the linked diagrams in the expression for $E_j^{(n)}$ is further clarified by examining the value of an unlinked diagram, e.g., the one given in Fig. 3.3A:

$$\mathrm{A} = (-1)^{1+1} \sum_{\mu p} \langle p|V|\mu\rangle\langle \mu|V|p\rangle \tfrac{1}{2}(-1)^{2+2} \sum_{\alpha\beta} \langle \alpha\beta||\alpha\beta\rangle \tag{3.60}$$

Because, for two noninteracting subsystems, both of the disjoint sums occurring in Eq. (3.60) are size consistent (i.e., proportional to the size of the system), the product would not be size consistent. Hence, unlinked diagrams correspond directly to non–size-consistent factors, which should not be included.

If H^0 is taken to be the HF Hamiltonian, so that $V = V_{\mathrm{HF}}$ of Eq. (3.33), then certain simplifications occur. In particular, all diagrams containing the loop structure cancel with corresponding diagrams having the potential symbol  in the same location except in first order. For example, diagrams B and C of Fig. 3.3 cancel since the value of diagram C is

$$\mathrm{C} = \tfrac{1}{2}(-1)^{2+2} \sum_{\substack{\alpha\beta \\ mnp}} \frac{\langle mp||\alpha\beta\rangle\langle \alpha\beta||np\rangle(-\langle n|V_{\mathrm{HF}}|m\rangle)}{(\varepsilon_\alpha + \varepsilon_\beta - \varepsilon_p - \varepsilon_n)(\varepsilon_\alpha + \varepsilon_\beta - \varepsilon_p - \varepsilon_m)} \tag{3.61}$$

which, because

$$\langle n|V_{\mathrm{HF}}|m\rangle = \sum_\mu \langle n\mu||m\mu\rangle \tag{3.62}$$

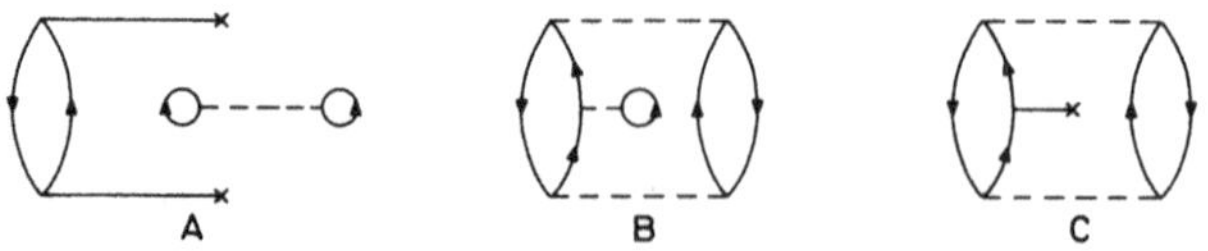

A B C

FIG. 3.3. Diagram A is unlinked and diagrams B and C demonstrate the cancellation of the Fock potential.

is equal to (but opposite in sign from) diagram B:

$$\mathrm{B} = \tfrac{1}{2}(-1)^{3+3} \sum_{\substack{\alpha\beta\mu \\ mnp}} \frac{\langle mp||\alpha\beta\rangle\langle n\mu||m\mu\rangle\langle\alpha\beta||np\rangle}{(\varepsilon_\alpha + \varepsilon_\beta - \varepsilon_p - \varepsilon_n)(\varepsilon_\alpha + \varepsilon_\beta - \varepsilon_p - \varepsilon_m)} \tag{3.63}$$

Because of the cancellation of  and  that occurs in all orders beyond first, we need only consider diagrams A, B, and C in Fig. 3.1 to determine the energy consistent through second order when $V = V_{\mathrm{HF}}$.

The first-order diagrams shown in Fig. 3.1A,B combine as follows:

$$\begin{aligned} \mathrm{A} + \mathrm{B} &= \tfrac{1}{2}(-1)^{2+2} \sum_{\alpha\beta} \langle\alpha\beta||\alpha\beta\rangle + (-1)^{1+1} \sum_{\alpha} (-)\langle\alpha|V_{\mathrm{HF}}|\alpha\rangle \\ &= -\tfrac{1}{2} \sum_{\alpha\beta} \langle\alpha\beta||\alpha\beta\rangle \end{aligned} \tag{3.64}$$

The second-order contribution may be written as

$$C = \sum_{\substack{\alpha\beta \\ rs}} (\tfrac{1}{2})^2 (-1)^{2+2} \frac{\langle\alpha\beta||rs\rangle\langle rs||\alpha\beta\rangle}{\varepsilon_\alpha + \varepsilon_\beta - \varepsilon_r - \varepsilon_s} \tag{3.65}$$

Both the first- and the second-order contributions are, of course, identical to the RSPT expression for $E_j^{(1)}$ and $E_j^{(2)}$ given in Eqs. (3.36) and (3.38), respectively.

In the application of MBPT to certain physical problems, it has been noticed that special families of diagrams seem to make important contributions to the energy in all orders. Attempts have been made to identify such diagrams and then to evaluate their energy contributions in a manner that permits these terms to be algebraically summed through all orders. One example of such a family of diagrams is shown in Fig. 3.4. We can apply the rules for Brandow diagrams given in Table II to each of these diagrams to obtain

$$(\mathrm{A}) = (\tfrac{1}{2})^2 \sum_{\substack{\alpha\beta \\ rs}} (-1)^{2+2} \frac{\langle\alpha\beta||rs\rangle\langle rs||\alpha\beta\rangle}{\varepsilon_\alpha + \varepsilon_\beta - \varepsilon_r - \varepsilon_s} \tag{3.66}$$

$$(\mathrm{B}) = (\tfrac{1}{2})^2 \sum_{\substack{\alpha\beta \\ rs}} (-1)^{2+4} \frac{\langle\alpha\beta||rs\rangle\langle\alpha\beta||\alpha\beta\rangle\langle rs||\alpha\beta\rangle}{(\varepsilon_\alpha + \varepsilon_\beta - \varepsilon_r - \varepsilon_s)^2} \tag{3.67}$$

$$(\mathrm{C}) = (\tfrac{1}{2})^2 \sum_{\substack{\alpha\beta \\ rs}} (-1)^{2+6} \frac{\langle\alpha\beta||rs\rangle\langle\alpha\beta||\alpha\beta\rangle^2\langle rs||\alpha\beta\rangle}{(\varepsilon_\alpha + \varepsilon_\beta - \varepsilon_r - \varepsilon_s)^3}, \tag{3.68}$$

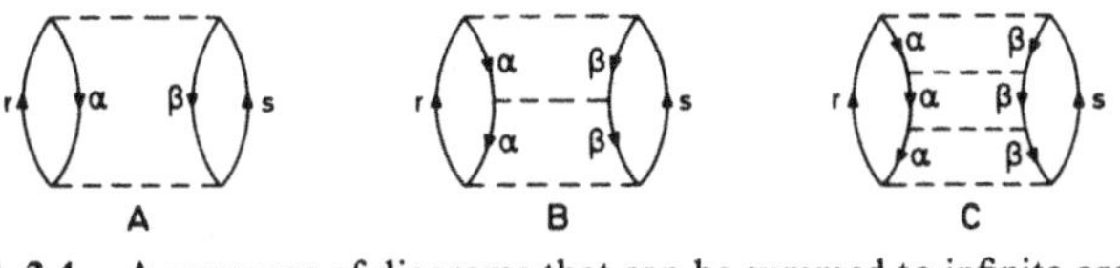

FIG. 3.4. A sequence of diagrams that can be summed to infinite order.

It should be clear that expression for higher-order diagrams analogous to Fig. 3.4A but with one or more additional interaction lines connecting α and β would involve higher powers of $\langle\alpha\beta||\alpha\beta\rangle(\varepsilon_\alpha + \varepsilon_\beta - \varepsilon_r - \varepsilon_s)^{-1}$. Therefore, the series represented by the sum of Fig. 3.4A–C is an easily summed geometric series of the form $y + yx + yx^2 + yx^3 + \cdots$. The result of summing this series is to yield

$$(\tfrac{1}{2})^2 \sum_{\substack{\alpha\beta\\ rs}} \langle\alpha\beta||rs\rangle\langle rs||\alpha\beta\rangle(\varepsilon_\alpha+\varepsilon_\beta-\varepsilon_r-\varepsilon_s)^{-1}[1-\langle\alpha\beta||\alpha\beta\rangle(\varepsilon_\alpha+\varepsilon_\beta-\varepsilon_r-\varepsilon_s)^{-1}]^{-1} \tag{3.69}$$

which can be rearranged to read

$$\frac{1}{4}\sum_{\substack{\alpha\beta\\ rs}} \langle\alpha\beta||rs\rangle\langle rs||\alpha\beta\rangle(\varepsilon_\alpha + \varepsilon_\beta - \varepsilon_r - \varepsilon_s - \langle\alpha\beta||\alpha\beta\rangle)^{-1} \tag{3.70}$$

We see that the result of summing the class of diagrams given in Fig. 3.4 is to generate an expression that is identical to the value of Fig. 3.4A except for the "denominator shift" of $-\langle\alpha\beta||\alpha\beta\rangle$.

In addition to expressing E_j in terms of diagrams, we may write the perturbation corrections to $|j\rangle$ in this language. A few first- and second-order Brandow wavefunction diagrams are shown in Figs. 3.5. The rules for evaluating these diagrams are similar to those for the energy except in two ways. First, in counting the number of hole lines to determine the sign $(-1)^h$, the external hole lines are not included. Second, each free external line has associated with it an excitation operator and an orbital energy term. For example, the value of diagram D in Fig. 3.5 is

$$\mathrm{D} = \tfrac{1}{2}(-1)^{1+2} \sum_{\substack{pq\\ \alpha\beta\gamma}} \frac{\langle pq||\alpha\beta\rangle\langle\alpha\beta||\gamma q\rangle}{(\varepsilon_\alpha + \varepsilon_\beta - \varepsilon_p - \varepsilon_q)(\varepsilon_\gamma - \varepsilon_p)} p^+\gamma|0^0\rangle \tag{3.71}$$

As was the case for diagram contributions to E_j, the factors $\succ\!\circ$ and $\succ\!\!-\!\!-\!\!\times$ that occur in equivalent locations, for example, in Figs. 3.5A,B, exactly cancel when V is V_{HF}. The cancellation in Fig. 3.5A,B is the diagrammatic expression of the BT; that is, the first-order wavefunction contains no singly excited configurations.

The kind of perturbation theories (RSPT and MBPT, in particular) described above have proven to be useful quantum-chemical tools. However, these methods are expected to fail whenever the perturbation (the electronic

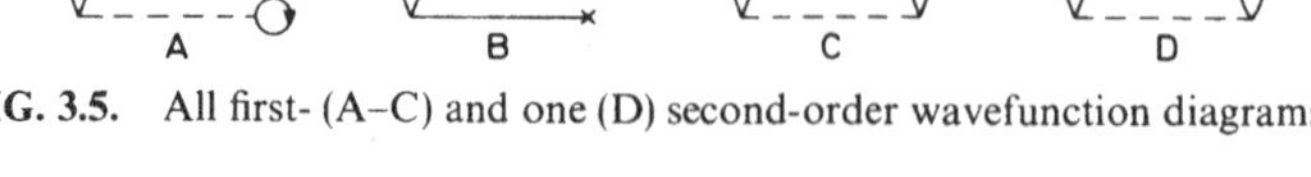

FIG. 3.5. All first- (A–C) and one (D) second-order wavefunction diagrams.

fluctuation potential if H^0 is the HF Hamiltonian) is not small or if the system under study is not well described in terms of a single Slater determinant. This would be the case, for example, when one or more of the states $|k^0\rangle$ is energetically close to $|j^0\rangle$ (i.e., $E_j^0 - E_k^0$ is small). This arises often when one breaks chemical bonds. Also, it is often not possible to describe a system in terms of a single determinant whose spin-orbital occupation is held constant throughout a large position of nuclear configuration space [e.g., $LiH(1\sigma^2 2\sigma^2)$ gives improper dissociation into ionic states at large R]. For these reasons, there has been recent research activity aimed at developing MBPT for a multiconfigurational reference state, but such tools are not yet commonly available. Thus although MBPT is indeed a size-consistent theory in that it yields total energies proportional to the size of the system for a collection of noninteracting subsystems, it still may suffer from the improper dissociation problem common to most single-configuration-based theories. Size consistency and proper dissociation are different characteristics. The question of proper dissociation has to do with whether the wavefunction contains configurations that can yield proper dissociation products and whether the method used to compute the amplitudes of these configurations (e.g., CI, MBPT, MCSCF) can be trusted to be accurate as dissociation occurs. The MBPT may indeed contain the $1\sigma^2 2\sigma 3\sigma$ configuration needed to dissociate LiH, but the amplitude of this configuration (which dominates at large R) cannot be obtained from an MBPT calculation based upon using the $1\sigma^2 2\sigma^2$ configuration as $|0^0\rangle$ unless a very high correlation level is considered. Hence although MBPT would yield a size-consistent energy for two or more noninteracting LiH molecules (each at their equilibrium bond lengths) it fails to describe even one LiH molecule at large bond lengths.

PROBLEMS

3.1 Using the orbital energies and two-electron integrals found in Problem 2.1, carry out a RSPT calculation of the first-order wavefunction $|1\sigma^2\rangle^{(1)}$ and the second-order energy $E^{(2)}$ for the case in which the zeroth-order wavefunction is taken to be the $1\sigma^2$ Slater determinant.

1. Show that the first-order wavefunction is given by

$$|1\sigma^2\rangle^{(1)} = -0.0442|2\sigma^2\rangle$$

2. Why does the $|1\sigma 2\sigma\rangle$ configuration not enter into the first-order wavefunction?

3. Normalize the resultant wavefunction that contains zeroth- plus first-order parts and compare it to the wavefunction obtained in the two-configuration CI study of Problem 2.3.

4. Show that the second-order RSPT correlation energy of HeH^+ is given by -0.0056 a.u. How does this compare with the correlation energy obtained from the two-configuration CI calculation?

5. Show that the second-order RSPT energy of a collection of n non-interacting HeH^+ ions reduces to n times the correlation energy of one such ion.

3.2 Using MBPT, determine the third-order contribution to the correlation energy for the HeH^+.

1. Write all Hugenholtz diagrams that contribute in third order when the perturbation $U = -V + W$ consists of the electronic repulsion W and a one-electron perturbation V.

In the following, assume now that $V = V_{HF}$.

2. Which of the diagrams of question 1 cancel?

3. Write the algebraic expression for the diagrams in question 2 that did not cancel.

The third-order contribution to the correlation energy for the HeH^+ of Problem 2.1 can now be determined, using the HF orbital energies and the one- and two-electron integrals in the HF basis that is determined there.

4. Determine the third-order contribution to the correlation energy for the HeH^+ system.

5. Compare the third-order contribution with the second-order contributions determined in Problem 3.1 and the full CI correlation energy.

SOLUTIONS

3.1

1. $$|j^{(1)}\rangle = -\sum_{\substack{\alpha<\beta \\ r<s}} \frac{\langle rs||\alpha\beta\rangle}{\varepsilon_r + \varepsilon_s - \varepsilon_\alpha - \varepsilon_\beta} r^+ s^+ \beta\alpha |j^0\rangle$$

In our case $\alpha = 1\sigma\alpha$, $\beta = 1\sigma\beta$, $r = 2\sigma\alpha$, $s = 2\sigma\beta$:

$$|1\sigma^{2(1)}\rangle = -\frac{\langle 22|11\rangle}{2(\varepsilon_2 - \varepsilon_1)}|2\sigma^2\rangle = -\frac{0.1261}{2[-0.2289 + 1.6562]}|2\sigma^2\rangle$$
$$= -0.0442|2\sigma^2\rangle$$

2. The BT gives $\langle 1\sigma^2|H|1\sigma 2\sigma\rangle = 0$; hence the $|1\sigma 2\sigma\rangle$ configuration does not enter into the first-order wavefunction.

3. $|0\rangle \cong |1\sigma^2\rangle - 0.0442|2\sigma^2\rangle$. To normalize, we divide by $[1 + (0.0442)^2]^{1/2} = 1.0010$:

$$|0\rangle = 0.9990|1\sigma^2\rangle - 0.0441|2\sigma^2\rangle$$

In the CI, we got

$$|0\rangle \cong 0.9984|1\sigma^2\rangle - 0.0556|2\sigma^2\rangle$$

4. $$E^{(2)} = -\frac{|\langle 22|11\rangle|^2}{2(\varepsilon_2 - \varepsilon_1)} = -\frac{(0.1261)^2}{2[-0.2289 + 1.6562]}$$
$$= -0.0056 \quad \text{a.u.}$$

From the two CI energy, compared to the SCF energy $\langle 1\sigma^2|H|1\sigma^2\rangle$, the correlation energy is $-4.2790 - (-4.2720) = -0.0070$ a.u.

5. $E^{(2)}$ is generally given by

$$\sum_{\substack{\alpha<\beta \\ m<p}} \frac{|\langle mp||\alpha\beta\rangle|^2}{\varepsilon_\alpha + \varepsilon_\beta - \varepsilon_m - \varepsilon_p}$$

For n noninteracting HeH^+ ions, the integrals $\langle mp||\alpha\beta\rangle$ involving orbitals on different ions vanish. Thus all four orbitals in $\langle mp||\alpha\beta\rangle$ must be on the same ion. Hence

$$E^{(2)} = \sum_{A=1} \left(\sum_{\substack{\alpha<\beta \\ m<p}} \frac{|\langle mp||\alpha\beta\rangle|^2}{\varepsilon_\alpha + \varepsilon_\beta - \varepsilon_m - \varepsilon_p} \right)_{\text{on } A}$$

where A runs over the n HeH^+ ions.

3.2

1. See Fig. 3.6, where the Hugenholtz diagrams are displayed.
2. Diagrams D1–X1 cancel with diagrams D2–X2 such that D1 cancels D2, E1 cancels E2, etc.
3. The Hugenholtz diagrams in Fig. 3.6A–C are translated into the corresponding Brandow diagrams in Fig. 3.7A–C, respectively. Each of these diagrams is then evaluated according to the rules to give

$$A = \sum_{\substack{mnpq \\ \alpha\beta}} \frac{1}{8} \frac{\langle\alpha\beta||mn\rangle\langle mn||pq\rangle\langle pq||\alpha\beta\rangle}{(\varepsilon_\alpha + \varepsilon_\beta - \varepsilon_m - \varepsilon_n)(\varepsilon_\alpha + \varepsilon_\beta - \varepsilon_p - \varepsilon_q)}$$

$$B = \sum_{\substack{\alpha\beta\gamma\delta \\ mn}} \frac{1}{8} \frac{\langle\alpha\beta||mn\rangle\langle\gamma\delta||\alpha\beta\rangle\langle mn||\gamma\delta\rangle}{(\varepsilon_\alpha + \varepsilon_\beta - \varepsilon_m - \varepsilon_n)(\varepsilon_\gamma + \varepsilon_\delta - \varepsilon_m - \varepsilon_n)}$$

$$C = \sum_{\substack{mnp \\ \alpha\beta\gamma}} \frac{\langle\alpha\gamma||mp\rangle\langle\beta m||n\alpha\rangle\langle np||\beta\gamma\rangle}{(\varepsilon_\alpha + \varepsilon_\gamma - \varepsilon_m - \varepsilon_p)(\varepsilon_\beta + \varepsilon_\gamma - \varepsilon_n - \varepsilon_p)}$$

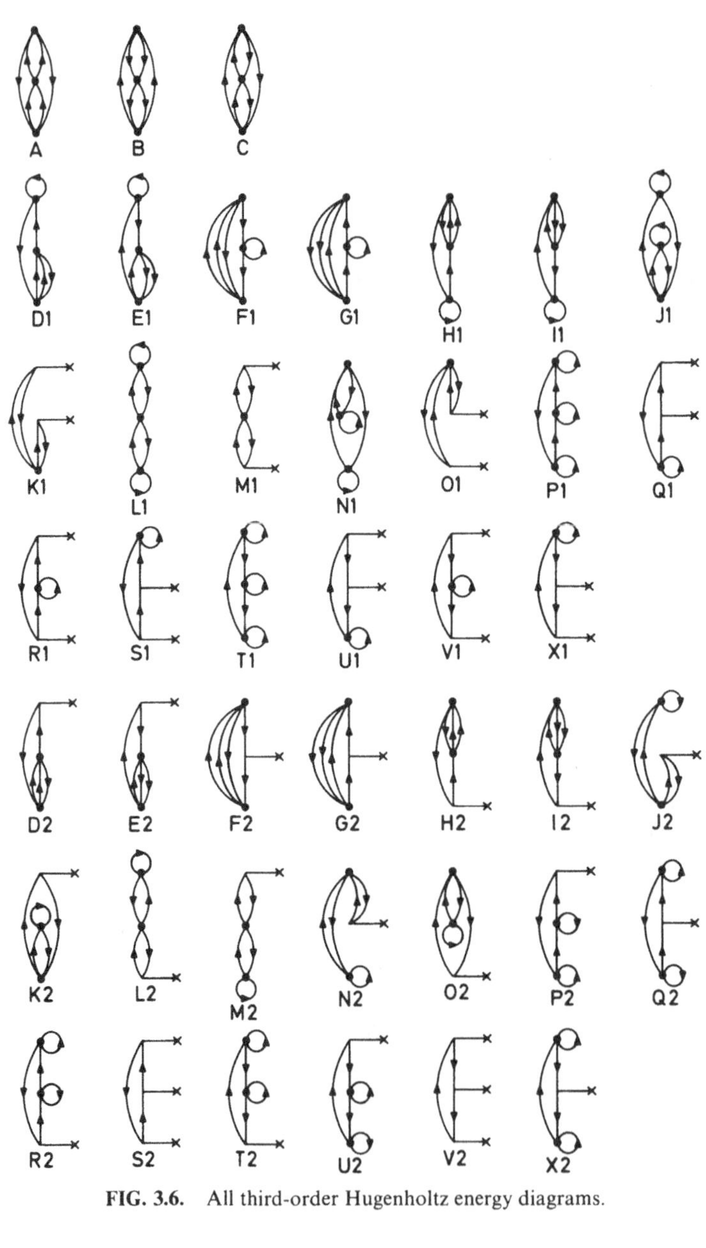

FIG. 3.6. All third-order Hugenholtz energy diagrams.

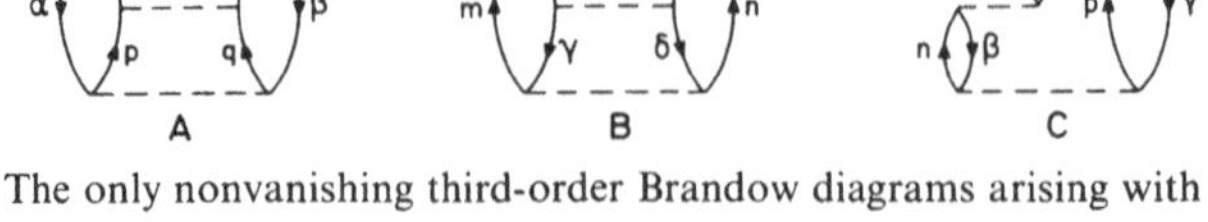

FIG. 3.7. The only nonvanishing third-order Brandow diagrams arising with HF orbitals.

4. $$A = \frac{\langle 11|22\rangle\langle 22|22\rangle\langle 22|11\rangle}{4(\varepsilon_1 - \varepsilon_2)^2} = 0.0012$$

$$B = \frac{\langle 11|22\rangle\langle 11|11\rangle\langle 22|11\rangle}{4(\varepsilon_1 - \varepsilon_2)^2} = 0.0019$$

$$C = \frac{\langle 11|22\rangle\langle 22|11\rangle(\langle 12|21\rangle - 2\langle 12|12\rangle)}{2(\varepsilon_1 - \varepsilon_2)^2} = -0.0042$$

$A + B + C = -0.0011$

5. Second order, -0.0056; second + third order, -0.0067; full CI -0.0072.

References

Bartlett, R. J., and Silver, D. M. (1975). *Int. J. Quantum Chem.* **S9**, 183.

Brandow, B. H. (1977). *Adv. Quantum Chem.* **10**, 187.

Brueckner, K. A. (1955a). *Phys. Rev.* **97**, 1353.

Brueckner, K. A. (1955b). *Phys. Rev.* **100**, 36.

Kelly, H. P. (1969). *Adv. Chem. Phys.* **14**, 129.

Linderberg, J., and Öhrn, Y. (1973). "Propagators in Quantum Chemistry." Academic Press, New York.

Löwdin, P. O. (1968). *Int. J. Quantum Chem.* **2**, 867.

March, N. H., Young, W. H., and Sampanthar, S. (1967). "The Many-Body Problem in Quantum Mechanics." Cambridge Univ. Press, Cambridge.

Pople, J. A., Seeger, R., and Krishnan, R. (1977). *Int. J. Quantum Chem.* **S11**, 149.

Raimes, S. (1972). "Many Electron Theory." North Holland Publ., Amsterdam.

Chapter 4 | The Coupled-Cluster Method

A. INTRODUCTION

The CI and MCSCF methods described earlier suffer from one significant weakness. The slow convergence of the wavefunction as the configuration size is increased is a problem that becomes more severe as the number of electrons in the system grows. In fact, for extended systems the finite CI or MCSCF wavefunctions (because they contain only a finite number of electron pair interactions) become infinitesimal portions of the exact wavefunction. Perturbation theory methods, whose wavefunction usually also contains only finite numbers of interactions, sometimes provide some relief because the total energy is not calculated as an expectation value. However, it is often not appropriate to assume that the usual fluctuation potential (true electron–electron interaction minus the HF potential) is small, i.e., to assume convergence of the perturbation series. Moreover, it is quite often important to be able to properly treat systems that are not adequately described by a single-configuration zeroth-order reference wavefunction (such as is assumed in most perturbation theories).

B. FORM OF THE WAVEFUNCTION

The coupled-cluster (CC) method (Cizek and Paldus, 1971; Harris, 1977a,b; Bartlett and Purvis, 1978) is an attempt to introduce interactions among electrons within clusters (predominantly pairs) as well as coupling among these clusters of electrons and to permit the wavefunction to contain all possible disjoint clusters. For example, we know, from the early work of Sinanoglu (1962) and others, that electron pair interactions are of utmost importance and that contributions of quadruply excited configurations to

$|0\rangle$ arise predominantly as products of doubly excited configurations. CC wavefunctions in which such electron pair interactions (clusters) are assumed to dominate still contain terms that describe disjoint products of electron pair clusters just as Sinanoğlu's observations would suggest. In fact, for a system containing an even (odd) number of electrons $N(N+1)$, one has products of 2, 3, . . . , $N/2$ disjoint pair clusters in the CC wavefunction. The mechanism for introducing these cluster interactions is to write the wavefunction $|0\rangle$ in terms of a so-called cluster operator T acting on a reference function describing noninteracting or noncoupled electrons $|0^0\rangle$:

$$|0\rangle = \exp(T)|0^0\rangle. \tag{4.1}$$

The reference function $|0^0\rangle$ has, in nearly all CC developments to date, been limited to a ket corresponding to a single Slater determinant. In the treatment given in this chapter, we therefore restrict our attention to this single determinantal case. The cluster operator T generates one-, two-electron, etc., clusters

$$T = T_1 + T_2 + \cdots + T_N \tag{4.2}$$

with

$$T_1 = \sum_{\alpha,r} t_\alpha^r r^+ \alpha \tag{4.3}$$

$$T_2 = \frac{1}{4} \sum_{\substack{\alpha\beta \\ rs}} t_{\alpha\beta}^{rs} r^+ s^+ \beta\alpha \tag{4.4}$$

etc. (the greek indices $\alpha, \beta, \gamma, \ldots$ denote spin-orbitals occupied in $|0^0\rangle$; $r, s, t, u, \ldots$ denote unoccupied spin-orbitals). To make some connection between the CC wavefunction of Eq. (4.1) and the more conventional CI and MBPT expressions for $|0\rangle$, we expand the $\exp(T)|0^0\rangle$ and collect terms of common excitation level:

$$\exp(T)|0^0\rangle = \Big(1 + T_1 + T_2 + \frac{1}{2!} T_1^2 + T_3 + \frac{1}{3!} T_1^3 + T_1 T_2 + T_4 + \frac{1}{4!} T_1^4 + \frac{1}{2!} T_2^2 + T_3 T_1 + \frac{1}{2!} T_1^2 T_2 + \cdots\Big)\Big|0^0\rangle \tag{4.5}$$

By grouping the terms of a given excitation level together, we see that the CC wavefunction can be rewritten as

$$\exp(T)|0^0\rangle = (1 + C_1 + C_2 + C_3 + \cdots)|0^0\rangle \tag{4.6}$$

where the configuration excitation operators $C_1, C_2, \ldots$ are

$$C_1 = T_1 \tag{4.7}$$

$$C_2 = T_2 + \frac{1}{2!} T_1^2 \tag{4.8}$$

$$C_3 = T_3 + \frac{1}{3!} T_1^3 + T_1 T_2 \tag{4.9}$$

$$C_4 = T_4 + \frac{1}{4!} T_1^4 + \frac{1}{2!} T_2^2 + T_3 T_1 + \frac{1}{2!} T_1^2 T_2 \tag{4.10}$$

etc. We thus see, for example, that the quadruple excitations that would be obtained in a CI or MCSCF treatment can be viewed within the cluster framework as consisting of five separate parts. The T_2^2 component is thought to be the dominant term because it represents the simultaneous interactions of two distinct pairs of electrons (e.g., electron pairs that occupy spatially different molecular orbitals). The T_4 term is usually expected to be quite small since it describes the simultaneous interaction of four electrons. The single-cluster contributions to C_4, C_3, and C_2 can be made small by using MCSCF orbitals.

If our normal description of chemical bonding in terms of electron pair bonds is correct, it is likely that a description of molecular structure in which T_2 is treated to high order [e.g., through $\exp(T_2)$] while T_1, T_3, T_4, etc. are either neglected or treated less rigorously, is quite accurate. For this reason we consider developing systematic procedures for truncating the expansion of T given in Eq. (4.2). By truncating [approximating the cluster operator T to some low-order (say pair clusters T_2)], the resultant wavefunction contains not only these low-order clusters $T_2|0^0\rangle$ but also disjoint clusters [e.g., $(1/2!)T_2T_2|0^0\rangle$, $(1/3!)T_2T_2T_2|0^0\rangle$, etc.] that involve more highly excited configurations than are present in $T_2|0^0\rangle$. Of course, these higher-order excitations [e.g., quadruply excited for $(1/2!)T_2T_2|0^0\rangle$] are present in $|0\rangle$ only to the extent that their amplitudes can be described in terms of products of the amplitudes belonging to the smaller clusters (e.g., $t_{\alpha\beta}^{rs}t_{\gamma\delta}^{tu}$). The fact that the product factors $T_2T_2|0^0\rangle$ contain only disjoint clusters arises because the operator product $(r^+s^+\cdots\alpha\beta\cdots)(t^+u^+\cdots\gamma\delta\cdots)$ vanishes if any of the hole $(\alpha, \beta, \gamma, \ldots)$ or particle $(r^+, s^+, t^+, \ldots)$ indices are equal. An essential point of the CC approach is that even low-order truncations of T (which are usually based upon the physical assumption that electron pair interactions dominate) lead to a wavefunction that contains all of the disjoint higher excitations needed to make the resultant energy (and other physical properties) size consistent.

C. EQUATIONS FOR THE CLUSTER AMPLITUDES

The cluster amplitudes $t_{\alpha\beta\cdots}^{rs\cdots}$ are determined by insisting that $\exp(T)|0^0\rangle$ satisfy the usual Schrödinger equation (or at least certain projections of this equation)

$$H\exp(T)|0^0\rangle = E\exp(T)|0^0\rangle \tag{4.11}$$

which upon premultiplying by $\exp(-T)$ gives

$$\exp(-T)H\exp(T)|0^0\rangle = E|0^0\rangle \tag{4.12}$$

The above exponential series gives, when expanded and collected together as commutators,

$$\left(H + [H,T] + \frac{1}{2!}[[H,T],T] + \frac{1}{3!}[[[H,T],T],T] + \frac{1}{4!}[[[[H,T],T],T],T]\right)\Big|0^\circ\rangle = E|0^\circ\rangle \tag{4.13}$$

The series truncates (exactly) after four commutators regardless of the level at which (T_n) T is truncated (if at all). This exact truncation is a result of the fact that H contains at most two-electron operators, which involve four general (particle or hole) operators i^+j^+lk. Therefore $[H,T]$ contains at most three general operators, $[[H,T],T]$ contains two, and $[[[[H,T],T],T],T]$ thus contains only (excitation) operators of the form $r^+s^+\cdots\alpha\beta\cdots$. These excitation operators clearly commute with T; thus the next (fifth) commutator in the series vanishes. The CC expression of the Schrödinger equation hence yields a quartic equation for the cluster amplitudes ($t_{\alpha\beta\cdots}^{rs\cdots}$) appearing in T.

A closed set of equations for the desired amplitudes is obtained by insisting that the final Schrödinger equation [Eq. (4.13)], when projected against a set of low-order excitations out of $|0^0\rangle$, yield zero. The particular excitations are usually chosen to include up through n-fold excitations from $|0^0\rangle$ in the case where T has been truncated at T_n. The resultant set of algebraic equations will then be equal in number to the number of amplitudes $t_{\alpha\beta\cdots}^{rs\cdots}$ in T. Once these amplitudes are obtained by solving the resultant nonlinear equations, the total electronic energy is computed by projecting Eq. (4.13) onto $|0^0\rangle$. We should stress that the energy expression thereby obtained is not variational in the sense that it is not given as an expectation value of the Hamiltonian. The quantity

$$\langle 0^0|\exp(T^+)H\exp(T)|0^0\rangle/\langle 0^0|\exp(T^+)\exp(T)|0^0\rangle$$

would give rise to a variational energy expression but the resultant commutator expansion of the exponential operators would not truncate because T^+ contains operators of the form $\alpha^+\beta^+ \cdots rs \cdots$, which do not commute with T. It is the presence of $\exp(-T)$, rather than $\exp(T^+)$, in Eq. (4.12) that gives rise to the exactly closed quartic equation for T. Moreover, the presence of the commutators in the expression for E and the fact that T contains only particle creation and hole annihilation operators makes the CC-calculated energy contain only linked terms (in the sense discussed in Chapter 3). This then makes E contain only size-consistent terms.

D. HARTREE–FOCK ORBITALS AND $T \simeq T_2$

Most CC calculations carried out so far have used the approximation $T \simeq T_2$. In this section, we treat this model in some detail since doing so will give us more insight into the structure of the CC equations. The physical motivation for approximating $T \simeq T_2$ relies on the fact that if the set of HF orbitals are used, the BT suggests that single excitation T_1 operators, which largely serve to optimize the spin-orbitals, should be less important than T_2. It is, however, now commonly felt that one should include both T_1 and T_2 so as to obtain a balanced or coupled description of the orbital and electron pair cluster optimization. Let us, however, continue our analysis of the $T \simeq T_2$ case.

To see what the solution of the above discussed nonlinear equations actually involves, let us examine these expressions in more detail for a case in which the spin-orbitals $\{\phi_\alpha\phi_\beta, \ldots, \phi_r, \ldots, \phi_s\}$ are eigenfunctions of a HF operator having orbital energies $\{\varepsilon_\alpha \cdots \varepsilon_r \cdots\}$. The decomposition of the Hamiltonian H into $H^0 + U$ is then given as in Eq. (2.84) by

$$H = H^0 + W - V_{\mathrm{HF}} \tag{4.14}$$

where H^0 is the HF Hamiltonian

$$H^0 = \sum_i \varepsilon_i i^+ i \tag{4.15}$$

W is the full electron interaction term in Eq. (2.84) and V_{HF} in the HF potential [Eq. (2.91)].

The commutator expansion of $\exp(-T)H\exp(T)$ in Eq. (4.12) given in Eq. (4.13) demonstrates in an elegant manner that when Eq. (4.13) is projected against low-order excitations $\langle{}^{rs\cdots n}_{\alpha\beta\cdots\gamma}| = \langle 0^0|\gamma \cdots \beta\alpha r^+ s^+ \cdots n^+$, it gives equations that are at most quartic in the cluster amplitudes $t^{rs\cdots u}_{\alpha\beta\cdots\gamma}$. However, it turns out that for finding equations for $t^{rs}_{\alpha\beta}$ it is equally simple to expand the exponential operators directly. To determine the total energy E, we project

Eq. (4.12) against $|0^0\rangle$ to obtain

$$\begin{aligned} E &= \langle 0^0| \exp(-T_2) H \exp(T_2) |0^0\rangle \\ &= \langle 0^0|(1 - T_2 + \tfrac{1}{2}(T_2)^2) H (1 + T_2 + \tfrac{1}{2}(T_2)^2 \cdots)|0^0\rangle \\ &= \langle 0^0|H|0^0\rangle + \langle 0^0|HT_2|0^0\rangle \\ &= E_{\text{HF}} + \sum_{\substack{m>n \\ \alpha>\beta}} \langle \beta\alpha| |mn\rangle t_{\alpha\beta}^{mn} \end{aligned} \tag{4.16}$$

where we have used the fact that $\langle 0^0|T_2 = 0$ because of the appearance of $r^+s^+\beta\alpha$ in T_2. We have also used the fact that $\langle 0^0|HT_2T_2|0^0\rangle$ vanishes because $T_2^2|0^0\rangle$ is quadruply excited and hence cannot couple through H to $\langle 0^0|$. The $t_{\alpha\beta}^{mn}$ amplitudes are determined by projecting Eq. (4.12) against doubly excited kets $\langle {}^{mn}_{\alpha\beta}|$ to obtain

$$0 = \langle {}^{mn}_{\alpha\beta}| \exp(-T_2) H \exp(T_2)|0^0\rangle \tag{4.17}$$

Expanding the exponential then allows one to see that the only nonvanishing contributions are contained in

$$0 = \langle {}^{mn}_{\alpha\beta}|H(1 + T_2 + \tfrac{1}{2}T_2^2)|0^0\rangle + \langle {}^{mn}_{\alpha\beta}|(-T_2)H(1 + T_2)|0^0\rangle \tag{4.18}$$

which shows that we obtain only a quadratic equation for the cluster amplitudes when $T \simeq T_2$. Explicitly evaluating the matrix element appearing in Eq. (4.18) then leads to the following nonlinear equation for the cluster amplitudes:

$$\begin{aligned} &(\varepsilon_m + \varepsilon_n - \varepsilon_\alpha - \varepsilon_\beta) t_{\alpha\beta}^{mn} \\ &\quad = \langle mn| |\alpha\beta\rangle - \sum_{p>q} \langle mn| |pq\rangle t_{\alpha\beta}^{pq} - \sum_{\gamma>\delta} \langle \gamma\delta| |\alpha\beta\rangle t_{\gamma\delta}^{mn} \\ &\qquad + \sum_{\gamma p} (\langle \gamma n| |\beta p\rangle t_{\alpha\gamma}^{mp} - \langle \gamma m| |\beta p\rangle t_{\alpha\gamma}^{np} - \langle \gamma n| |\alpha p\rangle t_{\beta\gamma}^{mp} + \langle \gamma m| |\alpha p\rangle t_{\beta\gamma}^{np}) \\ &\qquad + \sum_{\substack{\gamma>\delta \\ p>q}} \langle \gamma\delta| |pq\rangle [t_{\alpha\beta}^{pq} t_{\gamma\delta}^{mn} - 2(t_{\alpha\beta}^{mp} t_{\gamma\delta}^{nq} + t_{\alpha\beta}^{nq} t_{\gamma\delta}^{mp}) \\ &\qquad - 2(t_{\alpha\gamma}^{mn} t_{\beta\delta}^{pq} + t_{\alpha\gamma}^{pq} t_{\beta\delta}^{mn}) + 4(t_{\alpha\gamma}^{mp} t_{\beta\delta}^{nq} + t_{\alpha\gamma}^{nq} t_{\beta\delta}^{mp})] \end{aligned} \tag{4.19}$$

In the next sections we describe how solutions may be obtained to Eq. (4.19) and we discuss the relationship of the solution thereby obtained to results of MBPT.

E. PERTURBATIVE SOLUTION TO THE COUPLED-CLUSTER EQUATIONS

We describe here how Eq. (4.19) may be solved in a manner that shows the connection between the CC and the MBPT approaches. We solve

Eq. (4.19) in an iterative manner by performing an initial guess of $\{t_{\alpha\beta}^{rs}\}$ and then inserting this value on the right-hand side into Eq. (4.19) to obtain an improved set of values of $\{t_{\alpha\beta}^{rs}\}$. These are then inserted back into the right-hand side of Eq. (4.19) to again give us improved values of $\{t_{\alpha\beta}^{rs}\}$, etc. As an initial guess of the cluster amplitudes we set those amplitudes that appear on the right-hand side of Eq. (4.19) equal to zero. The motivation for this choice is that the terms containing t on the right-hand side of Eq. (4.19) are assumed to be smaller than those on the left-hand side of this equation. We then obtain the following expression for the amplitudes:

$$t_{\alpha\beta}^{mn} = \langle mn \| \alpha\beta \rangle (\varepsilon_m + \varepsilon_n - \varepsilon_\alpha - \varepsilon_\beta)^{-1} \tag{4.20}$$

Inserting this value of $t_{\alpha\beta}^{mn}$ into the CC expression for the total energy as given in Eq. (4.16) yields

$$E = E_{\mathrm{HF}} + \sum_{\substack{m>n \\ \alpha>\beta}} \langle \beta\alpha \| mn \rangle \langle mn \| \alpha\beta \rangle (\varepsilon_m + \varepsilon_n - \varepsilon_\alpha - \varepsilon_\beta)^{-1} \tag{4.21}$$

This expression is nothing but the result obtained in second-order perturbation theory, which is written explicitly in Eq. (3.38).

A second iteration may be carried out by inserting into the right-hand side of Eq. (4.19) the cluster amplitudes obtained above. If we then neglect the terms that are quadratic in the $t_{\alpha\beta}^{mn}$ amplitudes [the eighth through fourteenth terms on the right-hand side of Eq. (4.19)], we obtain cluster amplitudes that, when used to compute the energy E via Eq. (4.16) give the same algebraic expression as is obtained in third-order MBPT (see Problem 4.1, question 1). If these cluster amplitudes are then inserted into the right-hand side of Eq. (4.19) (keeping the quadratic terms this time), we obtain new amplitudes that, when used to compute E, give all contributions to the fourth-order MBPT energy that arise from quadruple excitations [C_4 in Eq. (4.10)]. From the form of our working equation, Eq. (4.19), it is further clear that the quadruple excitations obtained in this way can only arise from the $\frac{1}{2}(T_2)^2$ and $-T_2T_2$ terms. These terms, in a sense, correspond to two simultaneous interactions of two electrons (electron pair interaction). The T_4 term, which corresponds to a true four-body interaction, first enters at fifth order in perturbation theory, thus indicating that electron pair interactions are much more important than true four-body interaction (Sinanoğlu, 1962). All fourth-order energy diagrams can, of course, not be obtained by approximating T with T_2, since both single and triple excitations contribute in fourth order. To obtain all fourth-order diagrams in a CC calculation would require both T_1 and T_3 to be included in the cluster expansion.

The iterative process carried out when determining the cluster amplitudes from Eq. (4.19) may be continued by inserting the cluster amplitudes from one iteration into the right-hand side of Eq. (4.19) to obtain the new amplitudes. The converged result would then correspond to summing all diagrams that contain double and products of double excitations through infinite order.

F. NEWTON–RAPHSON METHOD

Clearly, either the equations obtained by taking $T \simeq T_2$ [Eq. (4.19)] or the general quartic equation obtained from Eq. (4.13) are nonlinear and multivariable. Such equations can be represented in matrix form (by defining $t_{\alpha\beta}^{rs}$ as the rs, $\alpha\beta$ element of the **t** column vector) as

$$\mathbf{0} = \mathbf{a} + \mathbf{bt} + \mathbf{ctt} \tag{4.22}$$

where, for example $a_{rs,\alpha\beta} = \langle rs||\alpha\beta\rangle$ [see Eq. (4.19)]. The solution of these nonlinear algebraic equations represents a substantial practical difficulty in implementing the CC method. To solve these equations one can employ the perturbative analysis described above. This technique has the advantages that it is straightforward to program on a computer and that it has a close connection with MBPT.

An alternative to the above described perturbative procedure is the multivariable Newton–Raphson method. Such methods were used in the first molecular CC calculations (Paldus *et al.*, 1972). Here, one attempts to choose **t** such that the vector **f**(**t**) defined as

$$\mathbf{f}(\mathbf{t}) \equiv \mathbf{a} + \mathbf{bt} + \mathbf{ctt} \tag{4.23}$$

becomes equal to zero. This is done by expanding **f**(**t**) about the "point" $\mathbf{t}_0$. Keeping only linear terms in this Taylor expansion and setting **f**(**t**) equal to zero, one obtains equations for the changes $\Delta\mathbf{t}$ in the **t** amplitudes, which can be expressed as

$$f_{\alpha\beta}^{rs}(\mathbf{t}) = 0 = f_{\alpha\beta}^{rs}(\mathbf{t}_0) + \sum_{\substack{uw\\ \gamma\delta}} \left(\frac{\partial f_{\alpha\beta}^{rs}}{\partial t_{\gamma\delta}^{uw}}\right)_{\mathbf{t}_0} \Delta t_{\gamma\delta}^{uw} \tag{4.24}$$

The step lengths (corrections to $\mathbf{t}_0$) can be obtained by solving the above set of linear equations and then used to update the **t** amplitudes

$$\mathbf{t} = \mathbf{t}_0 + \Delta\mathbf{t}. \tag{4.25}$$

These values of **t** can then be used as a new $\mathbf{t}_0$ vector for the next application of Eq. (4.24). This multidimensional Newton–Raphson procedure, which involves the solution of a large number of coupled linear equations, is then repeated until the $\Delta\mathbf{t}$ values are sufficiently small (convergence). Given the set of $t_{\alpha\beta}^{rs}$ amplitudes, Eq. (4.16) can then be used to compute E. Although the first applications of the coupled cluster method to quantum chemistry did employ this Newton–Raphson scheme, the numerical problems involved in solving the large multivariable inhomogenous equations [Eq. (4.24)] has led more recent workers to use the perturbative techniques discussed in Section E. Within the perturbative framework, more sophisticated methods have been developed to solve the large number of quadratic (for $T \simeq T_2$) equations that arise. One such device is based upon the so-called reduced linear equations technique, which has also been widely used to find selected eigenvalues and eigenvectors of large CI matrices (Davidson, 1975).

G. SUMMARY

Although the CC method possesses several advantages over CI and MCSCF approaches, the fact that the resultant set of CC equations that determine the $t_{\alpha\beta}^{rs} \cdots$ amplitudes are nonlinear and of very large dimensions even for modest-sized systems, has made the practical applications of this theory rather limited. An analysis of the relationships between the solutions of the nonlinear CC equations and the solutions of corresponding CI secular problems has recently been provided (Monkhorst and Zivkovic, 1978). This analysis thus provides some reason for optimism concerning the possibility of finding efficient mechanisms for solving the CC equations. However, at present, the nonlinear nature of the equations to be solved still makes the practical utilization of the CC method something toward which we are still working. Research aimed at achieving efficient solutions of the quadratic (or even quartic) coupled equations and at extending the CC development to open-shell and multiconfigurational reference states is necessary if the CC method is to become widely used in quantum chemistry.

PROBLEMS

4.1 Perform a CC calculation where T is approximated with T_2.

1. Show that the CC equations may be iterated to yield cluster amplitudes that, when used in the energy expression, give the MBPT third-order energy expression [see discussions following Eqs. (4.20) and (4.21)]. The third-order MBPT energy expression is given in Problem 3.2.

Carry out a CC calculation on HeH^+ using the minimum basis HF results found in Problem 2.1. In performing this calculation follow the steps given below.

2. First use a linear (truncated) form of the CC equation to determine the numerical values of the $t_{\alpha\beta}^{mn}$ parameters and then use these parameters to compute the corresponding correlation energy.

3. Argue that although the above linear form (question 2) of the CC equation and the perturbative solution (question 1) yield amplitudes that

have the same formal structure, the correlation energy of HeH^+ determined in question 2 and in third-order MBPT (question 1 and Problem 3.2) differ.

4. Use the quadratic form of the CC equations (which clearly has two solutions) to determine the values of the two sets of $t_{\alpha\beta}^{mn}$ parameters.

5. Evaluate the total energy and the correlation energy contribution for both of these two sets of solutions.

6. Show that the CC equations in Eqs. (4.16) and (4.18) and the CI eigenvalue equation that contains doubly excited states become identical for a two-basis-function two-electron problem. Why do the two configuration CI total energies of Problem 2.4 and the CC total energies of question 5 differ?

Consider now n HeH^+ molecules that are separated at infinite distance with each molecule described by the localized SCF orbitals of Problem 2.1.

7. Show by carrying out a perturbative solution to the CC equations as described in Section E that the correlation energy for the n HeH^+ molecules becomes identical to n times the correlation energy of a single HeH^+ molecule and that the CC model thus is size consistent.

SOLUTIONS

4.1

1. When the CC amplitudes on the right-hand side of Eq. (4.19) are set equal to zero, we get

$$t_{\alpha\beta}^{mn} = \langle mn||\alpha\beta\rangle(\varepsilon_m + \varepsilon_n - \varepsilon_\alpha - \varepsilon_\beta)^{-1}$$

Inserting this value of $t_{\alpha\beta}^{mn}$ on the right-hand side of Eq. (4.19) gives the next approximation to $t_{\alpha\beta}^{mn}$:

$$t_{\alpha\beta}^{mn} = (\varepsilon_m + \varepsilon_n - \varepsilon_\alpha - \varepsilon_\beta)^{-1}\Big[\langle mn||\alpha\beta\rangle - \sum_{p>q}\langle mn||pq\rangle t_{\alpha\beta}^{pq} - \sum_{\gamma>\delta}\langle\gamma\delta||\alpha\beta\rangle t_{\gamma\delta}^{mn}$$

$$+ \sum_{\gamma p}(\langle\gamma n||\beta p\rangle t_{\alpha\gamma}^{mp} - \langle\gamma m||\beta p\rangle t_{\alpha\gamma}^{np} - \langle\gamma n||\alpha p\rangle t_{\beta\gamma}^{mp} + \langle\gamma m||\alpha p\rangle t_{\beta\gamma}^{np})\Big]$$

The first term in the square brackets results in the second-order energy expression [Eq. (4.21)] when used in Eq. (4.16).

The second term in the square brackets gives, when inserted into Eq. (4.16), the correlation energy contribution

$$-\sum_{\substack{m>n\\ \beta>\alpha\\ p>q}} \frac{\langle\beta\alpha||mn\rangle\langle mn||pq\rangle\langle pq||\alpha\beta\rangle}{(\varepsilon_m + \varepsilon_n - \varepsilon_\alpha - \varepsilon_\beta)(\varepsilon_p + \varepsilon_q - \varepsilon_\alpha - \varepsilon_\beta)}$$

which is identical to diagram A in Fig. 3.7.

The third term gives

$$-\sum_{\substack{m>n\\ \alpha>\beta\\ \gamma>\delta}} \frac{\langle\beta\alpha||mn\rangle\langle\gamma\delta||\alpha\beta\rangle\langle mn||\gamma\delta\rangle}{(\varepsilon_m+\varepsilon_n-\varepsilon_\alpha-\varepsilon_\beta)(\varepsilon_m+\varepsilon_n-\varepsilon_\gamma-\varepsilon_\delta)}$$

which is identical to diagram B of Fig. 3.7.

The last four terms give

$$\sum_{\substack{m>n\\ \alpha>\beta}} \frac{\langle\beta\alpha||mn\rangle}{(\varepsilon_m+\varepsilon_n-\varepsilon_\alpha-\varepsilon_\beta)} \sum_{\gamma p}(\langle\gamma n||\beta p\rangle t_{\alpha\gamma}^{mp} - \langle\gamma m||\beta p\rangle t_{\alpha\gamma}^{np}$$

$$- \langle\gamma n||\alpha p\rangle t_{\beta\gamma}^{mp} + \langle\gamma m||\alpha p\rangle t_{\beta\gamma}^{np})$$

Substitution of variables [e.g., in the first term we substitute ($m\to p, \alpha\to\gamma, \gamma\to\beta, n\to m, \beta\to\alpha, p\to n$)] allows these four terms to be rewritten as

$$+\sum_{\substack{p>m\\ \gamma>\alpha\\ \beta n}} \frac{\langle\alpha\gamma||pm\rangle\langle\beta m||\alpha n\rangle\langle pn||\gamma\beta\rangle}{(\varepsilon_p+\varepsilon_m-\varepsilon_\alpha-\varepsilon_\gamma)(\varepsilon_p+\varepsilon_n-\varepsilon_\gamma-\varepsilon_\beta)}$$

$$-\sum_{\substack{m>p\\ \gamma>\alpha\\ n\beta}} \frac{\langle\alpha\gamma||mp\rangle\langle\beta m||\alpha n\rangle\langle pn||\gamma\beta\rangle}{(\varepsilon_m+\varepsilon_p-\varepsilon_\alpha-\varepsilon_\gamma)(\varepsilon_p+\varepsilon_n-\varepsilon_\gamma-\varepsilon_\beta)}$$

$$-\sum_{\substack{p>m\\ \alpha>\gamma\\ n\beta}} \frac{\langle\gamma\alpha||pm\rangle\langle\beta m||\alpha n\rangle\langle pn||\gamma\beta\rangle}{(\varepsilon_p+\varepsilon_m-\varepsilon_\gamma-\varepsilon_\alpha)(\varepsilon_p+\varepsilon_n-\varepsilon_\gamma-\varepsilon_\beta)}$$

$$+\sum_{\substack{m>p\\ \alpha>\gamma\\ n\beta}} \frac{\langle\gamma\alpha||mp\rangle\langle\beta m||\alpha n\rangle\langle pn||\gamma\beta\rangle}{(\varepsilon_m+\varepsilon_p-\varepsilon_\gamma-\varepsilon_\alpha)(\varepsilon_p+\varepsilon_n-\varepsilon_\gamma-\varepsilon_\beta)}$$

The above four terms when collected together give diagram C of Fig. 3.7. Hence, all second- and third-order diagrams have been accountered for.

2. The only nonvanishing cluster amplitude is $t_{1\alpha 1\beta}^{2\alpha 2\beta}$. Equation (4.19) gives

$$0=\langle 22|11\rangle + t_{1\alpha 1\beta}^{2\alpha 2\beta}(2\varepsilon_1 - 2\varepsilon_2 - \langle 22|22\rangle - \langle 11|11\rangle + 4\langle 12|12\rangle - 2\langle 12|21\rangle)$$

which gives $t_{1\alpha 1\beta}^{2\alpha 2\beta} = 0.0559$. Inserting this value in Eq. (4.16) gives the correlation energy contribution

$$\Delta E_{\text{corr}} = -0.0070 \quad \text{a.u.}$$

Notice that although $t_{1\alpha 1\beta}^{2\alpha 2\beta}$ is positive, the correlation energy of Eq. (4.16) is negative, because $\langle\beta\alpha||mn\rangle = -\langle 11|22\rangle$.

3. Both the third-order MBPT and the approximation in question 2 use a linear form of the CC equations. However in the MBPT solution (question 1) the cluster amplitudes used in Eq. (4.19) are determined from perturbation theory, which results in a second- and third-order correlation energy of -0.0066 (Solution 3.2, question 5). The nonperturbation solution of question 2 is different from the one obtained in question 1 and gives a correlation energy -0.0070.

4. The quadratic CC equation reads

$$0 = \langle 22|11\rangle + t_{1\alpha 1\beta}^{2\alpha 2\beta}(2\varepsilon_1 - 2\varepsilon_2 - \langle 22|22\rangle - \langle 11|11\rangle + 4\langle 12|12\rangle - 2\langle 12|21\rangle) + \langle 11|22\rangle t_{1\alpha 1\beta}^{2\alpha 2\beta} t_{1\alpha 1\beta}^{2\alpha 2\beta}$$

which gives

$$t_{1\alpha 1\beta}^{2\alpha 2\beta} = 0.0560, \qquad t_{1\alpha 1\beta}^{2\alpha 2\beta} = -17.8432$$

5. $t_{1\alpha 1\beta}^{2\alpha 2\beta} = 0.0560, \quad E = -4.2791, \quad E_{\text{corr}} = -0.0071$
$t_{1\alpha 1\beta}^{2\alpha 2\beta} = -17.8432, \quad E = -2.0220, \quad E_{\text{corr}} = 2.2500$

6. The CC (Schrödinger) Equation (4.12) contains only linear terms in T_2 when applied to a two-electron system:

$$(1 - T_2)H(1 + T_2)|0^0\rangle = E|0^0\rangle$$

When this equation is projected against $\langle 0^0|$ and $\langle {}^{2\alpha 2\beta}_{1\alpha 1\beta}| = \langle {}^{22}_{11}|$ one obtains

$$\langle 0^0|H|0^0\rangle + \langle 0|H|{}^{22}_{11}\rangle t = E \tag{A}$$

$$\langle {}^{22}_{11}|H|0^0\rangle + \langle {}^{22}_{11}|H|{}^{22}_{11}\rangle t - t\langle 0^0|H|0^0\rangle - t^2\langle 0|H|{}^{22}_{11}\rangle = 0 \tag{B}$$

where $t_{1\alpha 1\beta}^{2\alpha 2\beta}$ is denoted t. Substituting Eq. (A) into Eq. (B) gives

$$\langle {}^{22}_{11}|H|0^0\rangle + \langle {}^{22}_{11}|H|{}^{22}_{11}\rangle t = Et \tag{C}$$

Equations (A) and (C) are nothing but the CI eigenvalue problem written out in component form for an intermediate normalized eigenvector with components $(1, t)$. The CC total energies $(-4.2791, -2.0220)$ and the CI total energies $(-4.2790, -2.0079)$ differ only because of numerical errors caused by using four significant digits in the integrals.

7. When the cluster amplitudes in the nonlinear part of Eq. (4.19) are set equal to zero (as in the first step of the perturbative solution), the only nonvanishing cluster amplitudes that remain (see the solution to question 1) are those involving all four orbitals in $t_{\alpha\beta}^{mn}$ located on the same HeH^+ molecule. This result is due to the fact that integrals involving orbitals on different HeH^+ molecules are zero. Continuing this iterative process does not introduce cluster amplitudes that couple different HeH^+ molecules again because integrals involving orbitals on two or more different molecules vanish. Hence,

the CC equations separate into equations for each HeH^+ molecule. Consequently, the correlation energy as computed via Eq. (4.16) for n HeH^+ molecules will be n times the contribution from a single HeH^+ molecule.

References

Bartlett, R. J., and Purvis, G. D. (1978). *Int. J. Quantum Chem.* **16**, 561.
Cizek, J., and Paldus, J. (1971). *Int. J. Quantum Chem.* **5**, 359.
Davidson, E. R. (1975). *J. Comput. Phys.* **17**, 87.
Harris, F. E. (1977a). *Int. J. Quantum Chem.* **S11**, 403.
Harris, F. E. (1977b). "Electrons in Finite and Infinite Structures" (P. Phariseau and L. Scheive, eds.). Plenum, New York.
Monkhorst, H., and Zivković, T. (1978). *J. Math. Phys.* **19**, 1007.
Paldus, J., Cizek, J., and Shavitt, I. (1972). *Phys. Rev. A* **5**, 50.
Sinanoğlu, O. (1962). *J. Chem. Phys.* **36**, 706.

Chapter 5 | Physical Properties

A. CLASSES OF PROPERTIES

Given wavefunctions belonging to one or more states that are obtained from an MCSCF, HF, CI, RSPT, or CC calculation, one is often interested in subsequently using these wavefunctions to compute physical properties of the system other than the total electronic energy. Below we discuss how the three distinct classes of properties—expectation values, transition properties, and response properties—may be evaluated, and we show also how stationary points on the potential energy surface may be determined using a quadratically convergent procedure.

1. Expectation Values

State average values such as dipole and quadrupole moments and electron spin densities are usually evaluated as expectation values of their corresponding quantum-mechanical operators. For example, the electronic contribution to the dipole moment operator is

$$\mathbf{r} = \sum_{i,j} e\langle\phi_i|\mathbf{r}|\phi_j\rangle i^+ j \tag{5.1}$$

and the electronic contribution to the dipole moment of state $|0\rangle$ thus becomes $-\langle 0|\mathbf{r}|0\rangle$. In evaluating expectation values, we must be careful that the wavefunction being used is of sufficiently high quality to permit accurate results. For example, in computing the expectation value of the electronic contributions to the dipole moment beyond the SCF level, it is important to include singly excited configurations in the CI or MCSCF wavefunction. A perturbation analysis of the order in which singly excited, doubly excited, etc. configurations enter in the calculation of the dipole moment makes this statement easily understood. The first-order RSPT function, which includes only doubly excited configurations (relative to the single determinantal zeroth-order function $|0^0\rangle$), yields a dipole moment average value that differs from $\langle 0^0|\mathbf{r}|0^0\rangle$ only in second order since $\langle 0^0|\mathbf{r}|^{rs}_{\alpha\beta}\rangle$ vanishes as a result of the fact that $\mathbf{r}$ is a one-electron operator. As both the singly and doubly excited states contribute in second order, it becomes equally important to include both singly and doubly excited configurations in the calculation. This observation demonstrates the point that those configurations that are optimal for describing the total electronic energy may not be adequate for obtaining accurate expectation values. This conclusion is now generally accepted as applying to all types (MCSCF, CI, HF, CC) of wavefunctions and is important to keep in mind when choosing which configurations to employ in any calculation.

Within the class of expectation values, we might also include calculations of electronic excitation and ionization energies as differences in individual state energies. The excitation and ionization energies are small numbers compared to the individual state total energies. For this reason, alternative procedures have been developed that can be employed to directly calculate such excitation (and ionization) energies as well as their corresponding oscillator strengths and that avoid the difficulties that might appear when subtracting two large numbers the difference of which is a small number. These direct evaluation techniques are based upon the so-called Green's function (GF) methods described in Chapter 6. The energy differences obtained either from a GF or by subtracting two wavefunction expectation values no longer have the upper bound property that individual state energies possess. Thus, there is no fundamental reason to insist that excitation energies be calculated as differences between state expectation values each of which are upper bounds to two state energies.

2. Transition Properties

The second class of quantities in which one is likely to be interested we refer to as transition properties. They include, for example, the electric dipole transition moment $\langle 0|\mathbf{r}|n\rangle$ between stationary states $|0\rangle$ and $|n\rangle$. The primary difficulty in evaluating such transition moments has to do with treating the overlap between nonorthogonal orbitals that arises in computing $\langle 0|i^+j|n\rangle$. That is, unless $|n\rangle$ and $|0\rangle$ are both expressed as linear combinations of determinants involving a common set of orthonormal spin-orbitals, the determinants in $|n\rangle$ will not be orthogonal to those in $|0\rangle$. Rather than being an exceptionally rare situation, this is actually the most likely case. For example, MCSCF calculations or INO-CI calculations on two electronic states of a molecule invariably result in different optimal (MCSCF or INO) orbitals for the two states. Although these nonorthogonality problems do indeed make the evaluation of transition properties quite difficult, it is still possible to compute the requisite overlap matrices and thereby obtain the desired quantity. However, this formidable difficulty provides strong motivation for us to treat the evaluation of transition moments through the GF framework as described in Chapter 6. Suffice it to say for now that these GF methods are designed to yield both transition moments and electronic energy differences directly rather than as matrix elements and energy differences of two separate states.

3. Response Properties

In addition to expectation values and transition moments, we have a third class of important physical properties, which we refer to as second-order response properties. To develop some understanding for the meaning of and theoretical methods for studying these responses, let us investigate the response of a state $|0\rangle$ corresponding to H to an external time-independent one-electron perturbation (αH_1)

$$H \rightarrow H + \alpha H_1 \tag{5.2}$$

Such perturbations could, for example, include electric field ($\alpha = \mathscr{E}$) effects or nuclear coordinate displacements. The total electronic energy in the presence of the perturbation becomes a function of α and may (for small α) be expanded in a power series

$$E(\alpha) = \langle\tilde{0}|H + \alpha H_1|\tilde{0}\rangle = E_0 - \alpha E_1 - \tfrac{1}{2}\alpha^2 E_2 - \tfrac{1}{6}\alpha^3 E_3 - \tfrac{1}{24}\alpha^4 E_4 \ldots \tag{5.3}$$

The terms that are nonlinear in α arise because the state wavefunction $|\tilde{0}\rangle$ depends on α (i.e., the state has responded to αH_1, which gives rise to the name "response property"). When, for example, αH_1 represents a static electric field ($\alpha H_1 = \mathscr{E} \cdot \mathbf{r}$), E_1 yields the permanent electric dipole moment (μ) of the unperturbed state $|0\rangle$, E_2 gives this state's polarizability ($\boldsymbol{\alpha}$), and E_3, E_4, etc. yield successively higher hyperpolarizabilities (β, γ, etc.).

a. *Finite-Field Approach*

One way of determining the first- and second-order response properties would be to calculate the total electronic energy of the system with αH_1 present (using the CI, HF, RSPT, MCSCF, or CC method) for several small values of α and to then attempt to fit these computed $E(\alpha)$ values to the series given in Eq. (5.3). This numerical procedure is usually referred to as the finite-field method. As an alternative to performing a least-squares fit to Eq. (5.3) one may, by judiciously choosing the values of the field at which $E(\alpha)$ is computed, employ versions of Eq. (5.3) that contain only odd or even powers of α $[E_\pm = E(\alpha) \pm E(-\alpha)]$. Furthermore, by combining computed values of $E_\pm(\alpha)$ and $E_\pm(2\alpha)$, one can selectively remove higher (odd or even) powers of α from the resultant equation. For example, by using $-\tfrac{2}{3}E_-(\alpha) + \tfrac{1}{12}E_-(2\alpha)$ one obtains $E_1\alpha + O(\alpha^5)$ since the $E_3\alpha^3$ was cancelled by taking the proper $(-\frac{2}{3}, \frac{1}{12})$ combination of $E_-(\alpha)$ and $E_-(2\alpha)$. Alternatively, using $E_+(\alpha)$ and $E_+(2\alpha)$ one can obtain

$$-\tfrac{4}{3}E_+(\alpha) + \tfrac{1}{12}E_+(2\alpha) + 5E_0 + O(\alpha^6) = \alpha^2 E_2 \tag{5.4}$$

Again, notice the cancellation of the power of α two higher than the power occurring in the property being evaluated. These finite-difference fits (Bartlett and Purvis, 1979) of Eq. (5.3) to calculated values of $E(\alpha)$ then permit one to obtain the dipole moment μ from E_1 and the polarizability from E_2.

Although the numerical procedure outlined above may permit one to efficiently and precisely extract from computed energy values $[E(\alpha)]$ the desired response properties, it by no means guarantees the accuracy of these properties. The accuracy of the computed response properties is determined by the quality of the wavefunction $|\tilde{0}\rangle$ used to evaluate $E(\alpha)$. It is not at all straightforward to choose an atomic basis set that permits the orthonormal molecular orbitals appearing in $|\tilde{0}\rangle$ to properly polarize in the presence of the field. Furthermore, it is difficult to choose a set of configurations for use in constructing $|\tilde{0}\rangle$ that is certain to yield the same accuracy in the computed $E(\alpha)$ values for all values of the field strengths α. Because of these difficulties, it is important to look for alternative methods for computing response properties. In Section B, we outline an analytical approach to this problem that does not involve fitting values of the energy that are computed at finite values of the applied-field strength.

b. Analytical Approach

As an alternative that does not suffer from these difficulties, analytical expressions for the response properties may be derived. If we are able to obtain a closed-form expression for the response of a state wavefunction $|0\rangle$ to the presence of the "field" αH_1,

$$|\tilde{0}\rangle = A^{-1/2}[|0\rangle + \alpha|0^1\rangle + \alpha^2|0^2\rangle + \cdots] \tag{5.5}$$

(A is a normalization constant), then this result can be used in Eq. (5.3) to express the Hamiltonian expectation value $\langle\tilde{0}|H + \alpha H_1|\tilde{0}\rangle$ as a power series in α, upon which the desired second-order response is identified as the multiplier of α^2. Of course, for each specific choice of the form of $|0\rangle$ (i.e., MCSCF, CI, RSPT, CC) the prescription for evaluating Eq. (5.5) is different; the basic approach is, however, identical for all such wavefunctions.

B. MCSCF TREATMENT OF RESPONSE

To illustrate the analytical approach, let us consider how an MCSCF wavefunction would respond to a one-electron external perturbation of the form

$$H \to H + \alpha H_1 + \alpha^2 H_2 \tag{5.6}$$

The total energy of the system in the presence of the external field may be written as

$$\begin{aligned} E(\alpha, \lambda, S) &= \langle 0|\exp(-iS)\exp(-i\lambda)(H + \alpha H_1 + \alpha^2 H_2)\exp(i\lambda)\exp(iS)|0\rangle \\ &= \langle 0|H + \alpha H_1 + \alpha^2 H_2|0\rangle - i\langle 0|[S + \lambda, H + \alpha H_1]|0\rangle \\ &\quad + \tfrac{1}{2}\langle 0|[S,[H,S]]|0\rangle + \tfrac{1}{2}\langle 0|[\lambda,[H,\lambda]]|0\rangle \\ &\quad + \langle 0|[S,[H,\lambda]]|0\rangle + \ldots \end{aligned} \tag{5.7}$$

The values of **P** and **κ** appearing in S and λ, respectively, may be expanded as power series in α:

$$S = S^{(0)} + \alpha S^{(1)} + \alpha^2 S^{(2)} + \cdots \tag{5.8}$$

$$\lambda = \lambda^{(0)} + \alpha\lambda^{(1)} + \alpha^2\lambda^{(2)} + \cdots \tag{5.9}$$

Since the λ and S operators are determined by making the total energy expression in Eq. (5.7) stationary, the zeroth-order terms that appear in Eqs. (5.8) and (5.9) become zero because the state $|0\rangle$ was optimized in the absence of the one-electron perturbation. The terms $-i\langle 0|[\lambda^{(1)} + S^{(1)}, H]|0\rangle$, which are of first order in α, and $-i\langle 0|[\lambda^{(2)} + S^{(2)}, H]|0\rangle$, which are of second order, vanish because of the GBT. Hence, in Eq. (5.7) all of the terms remaining should be viewed as containing $\lambda^{(1)}$ and $S^{(1)}$ since we are only keeping terms up through α^2 in our energy expansion.

Using Eqs. (2.29) and (2.30), we may express the above total energy in a form similar to the one given in Eq. (2.25):

$$\begin{aligned} E(\alpha, \lambda, S) &= E(0,0,0) + \alpha\langle 0|H_1|0\rangle + \alpha^2\langle 0|H_2|0\rangle \\ &\quad - \alpha 2(\boldsymbol{\kappa}\mathbf{P})\begin{pmatrix}\mathbf{F}\\ \mathbf{G}\end{pmatrix} + (\boldsymbol{\kappa}\mathbf{P})(\mathsf{A} - \mathsf{B})\begin{pmatrix}\boldsymbol{\kappa}\\ \mathbf{P}\end{pmatrix} + \cdots \end{aligned} \tag{5.10}$$

where the matrices **F** and **G** are defined as

$$\mathbf{F} = \langle 0|[Q, H_1]|0\rangle \tag{5.11}$$

$$\mathbf{G} = \langle 0|[R, H_1]|0\rangle \tag{5.12}$$

and Q and R are given in Eq. (2.26).

Since the total energy must be stationary in the presence of the external perturbation, we may determine **κ** and **P** from Eq. (5.10). Neglecting third- and higher-order terms, we obtain by differentiating with respect to **κ** and **P**

$$-2\alpha\begin{pmatrix}\mathbf{F}\\ \mathbf{G}\end{pmatrix} + 2(\mathsf{A} - \mathsf{B})\begin{pmatrix}\boldsymbol{\kappa}\\ \mathbf{P}\end{pmatrix} = 0 \tag{5.13}$$

which may be written as

$$\begin{pmatrix}\boldsymbol{\kappa}\\ \mathbf{P}\end{pmatrix} = \alpha(\mathsf{A} - \mathsf{B})^{-1}\begin{pmatrix}\mathbf{F}\\ \mathbf{G}\end{pmatrix} \tag{5.14}$$

Note that the $\mathsf{A} - \mathsf{B}$ matrix is evaluated from Eqs. (2.29) and (2.30) and does not contain H_1 or H_2. Using Eq. (5.14) to eliminate $\boldsymbol{\kappa}$ and $\mathbf{P}$ in Eq. (5.10), we obtain an expansion of the total energy as a function of α:

$$E(\alpha) = \langle 0|H|0\rangle + \alpha\langle 0|H_1|0\rangle + \alpha^2\langle 0|H_2|0\rangle$$
$$- \alpha^2(\mathbf{FG})(\mathsf{A} - \mathsf{B})^{-1}\begin{pmatrix}\mathbf{F}\\ \mathbf{G}\end{pmatrix} + O(\alpha^3) \quad (5.15)$$

Notice that in this MCSCF result, the multiplier of α is equal to the expectation value of the perturbation operator H_1. We have thus obtained an analytical expression from which to determine the desired first- and second- order response properties. This analytical approach for determining the second-order properties is referred to as the coupled multiconfiguration Hartree–Fock (CMCHF) approach (Dalgaard and Jørgensen, 1978).

If only one configuration is used for expanding the reference state $|0\rangle$ the above development can still be used to give

$$E(\alpha) = \langle 0|H|0\rangle + \alpha\langle 0|H_1|0\rangle + \alpha^2\langle 0|H_2|0\rangle$$
$$- \alpha^2\mathbf{F}(\mathsf{A}_{11} - \mathsf{B}_{11})^{-1}\mathbf{F} + O(\alpha^3) \quad (5.16)$$

where A_{11} and B_{11} are defined in Eqs. (2.29) and (2.30). This approximation to second-order properties has been denoted the coupled Hartree–Fock (CHF) method.

C. CI RESPONSE PROPERTIES

In a CI approach to this same problem, the variation of the reference state is described through variations in the configuration expansion coefficients. These variations may be described either by the $\exp(iS)$ operator or through the linear variational parameters C_{g0}. Because orbital variations are not considered in such a CI calculation, first- and second-order properties may be easily determined from Eq. (5.15) by neglecting all terms that involve the orbital optimization parameter $\boldsymbol{\kappa}$:

$$E(\alpha) = \langle 0|H|0\rangle + \alpha\langle 0|H_1|0\rangle - \alpha^2\mathbf{G}\mathsf{A}_{22}^{-1}\mathbf{G} + O(\alpha^3) + \alpha^2\langle 0|H_2|0\rangle \quad (5.17)$$

where

$$G_n = \langle n|H_1|0\rangle \quad (5.18)$$

and

$$(A_{22})_{mn} = \langle m|H|n\rangle - \delta_{mn}\langle 0|H|0\rangle \quad (5.19)$$

The matrix A_{22} contains the CI matrix involving all states $|n\rangle$ except the reference state $|0\rangle$. Carrying out a CI calculation (with H not including αH_1) within this orthogonal complement space would lead to the following familiar expression for the second-order property:

$$\mathbf{G}\mathbf{A}_{22}^{-1}\mathbf{G} = \sum_{n \neq 0} |\langle 0|H_1|n\rangle|^2 (E_n - E_0)^{-1} \tag{5.20}$$

However, this diagonalization of $\langle n|H|m\rangle$ is not necessary; Eq. (5.17) still gives the CI approximation to the desired second-order property. It should be noted that the second-order properties obtained from Eq. (5.20) simulate a finite-field CI calculation where the same orbitals are used to obtain the total energy at various strengths of the field. Hence, it is appropriate to take $\boldsymbol{\kappa} = \mathbf{0}$ in deriving Eq. (5.20) because the orbitals used have not been determined in the presence of the external field.

A finite-field CI calculation in which the orthonormal orbitals used to construct $|0\rangle$ are determined via an SCF calculation in the presence of the applied field could not easily be described in the analytical framework given here. Taking $\boldsymbol{\kappa} = \mathbf{0}$ is not appropriate because the orbitals are "optimized" with the field present. However, the orbitals are determined from a single-configuration (SCF) calculation rather than through the simultaneous optimization of $\boldsymbol{\kappa}$ and $\mathbf{P}$ for a multiconfiguration wavefunction. Hence a significant disadvantage of such a finite-field CI method is that it can not easily be directly connected with the analytical response equation given earlier.

D. THE HELLMANN–FEYNMAN THEOREM

It follows from the above MCSCF-based derivation that the Hellmann–Feynman theorem is fulfilled both for SCF and MCSCF wavefunctions since Eq. (5.15) yields, upon differentiation with respect to α,

$$\left.\frac{dE(\alpha)}{d\alpha}\right|_{\alpha=0} = \langle 0|H_1|0\rangle \tag{5.21}$$

It should, however, be pointed out that this result is a consequence of the fact that the SCF and MCSCF wavefunctions $|0\rangle$ have been optimized with respect to all variational parameters in $|0\rangle$ and that $\lambda^{(0)}$ and $S^{(0)}$ in Eqs. (5.8) and (5.9) therefore vanish. If the orbital optimization is carried out using a limited number of the total set of variational parameters in $|0\rangle$, the expansions in Eqs. (5.8) and (5.9) contain zeroth-order elements. The expansion of the total energy $E(\alpha)$ would then contain first-order terms in α beyond $\langle 0|H_1|0\rangle$ and the Hellmann–Feynman theorem would therefore not be fulfilled. This is the case in a limited CI calculation where the orbital variations are not considered explicitly [Eq. (5.9) contains zeroth-order terms]. Of course, the Hellmann–Feynman theorem is fulfilled in the full CI limit, where the orbital optimization parameters are redundant.

E. THE COUPLED-CLUSTER RESPONSE PROPERTIES

The above described linear-response approach to calculating first- and second-order properties can also be applied within the CC and RSPT frameworks. In the former (Monkhorst, 1977) theory we consider the CC working equations for a Hamiltonian to which a one-electron perturbation αH_1 has been added:

$$H(\alpha) = H + \alpha H_1 \tag{5.22}$$

$$\exp[-T(\alpha)]H(\alpha)\exp[T(\alpha)]|0^0\rangle = E(\alpha)|0^0\rangle \tag{5.23}$$

Neglecting the variations in the orbitals when the field is applied, the equation for E and the cluster amplitudes $t_{\alpha\beta\cdots}^{rs\cdots}$,

$$\langle 0^0|\exp[-T(\alpha)]H(\alpha)\exp[T(\alpha)]|0^0\rangle = E(\alpha) \tag{5.24}$$

and

$$\langle_{\alpha\beta\cdots}^{rs\cdots}|\exp[-T(\alpha)]H(\alpha)\exp[T(\alpha)]|0^0\rangle = 0 \tag{5.25}$$

can be expanded in powers of the field α once the cluster operators $T(\alpha)$ and $E(\alpha)$ are so expanded:

$$T(\alpha) = T^0 + aT^{(1)} + \alpha^2 T^{(2)} + \cdots \tag{5.26}$$

$$E(\alpha) = E^0 + \alpha E^{(1)} + \alpha^2 E^{(2)} + \cdots \tag{5.27}$$

The resultant first- and second-order equations read

$$E^{(1)} = \langle 0^0|\exp(-T^0)\{H_1 + [H, T^{(1)}]\}\exp(T^0)|0^0\rangle \tag{5.28}$$

$$0 = \langle_{\alpha\beta\cdots}^{rs\cdots}|\exp(-T^0)\{H_1 + [H, T^{(1)}]\}\exp(T^0)|0^0\rangle \tag{5.29}$$

and

$$E^{(2)} = \langle 0^0|\exp(-T^0)\{[H_1, T^{(1)}] + \tfrac{1}{2}[[H, T^{(1)}], T^{(1)}] + [H, T^{(2)}]\}\exp(T^0)|0^0\rangle \tag{5.30}$$

$$0 = \langle_{\alpha\beta\cdots}^{rs\cdots}|\exp(-T^0)\{[H_1, T^{(1)}] + \tfrac{1}{2}[[H, T^{(1)}], T^{(1)}] + [H, T^{(2)}]\}\exp(T^0)|0^0\rangle \tag{5.31}$$

respectively. The zeroth-order (in α) equations are, of course, nothing but the original CC equations in the absence of αH_1. We assume that we have already solved these equations. It is probably most reasonable to choose $T^{(1)}$ and $T^{(2)}$ to contain the operators $r^+s^+\cdots\alpha\beta\cdots$, which are of no higher cluster size than those in T^0 (e.g., $T^0 = T_1 + T_2$ is quite likely to be chosen for physical and practical reasons).

The above first-order equation for $T^{(1)}$ [Eq. (5.29)] expresses a set of linear algebraic equations for the cluster amplitudes in $T^{(1)}$, which can be written in matrix form as

$$\mathsf{D}\mathbf{t}^{(1)} = \mathbf{H}_1 \tag{5.32}$$

where the elements of $\mathbf{H}_1$ and D, respectively, are

$$-\langle{}^{rs\cdots}_{\alpha\beta\cdots}|\exp(-T^0)H_1\exp(T^0)|0^0\rangle,$$

$$\langle{}^{rs\cdots}_{\alpha\beta\cdots}|\exp(-T^0)[H, p^+q^+\cdots\gamma\delta\cdots]\exp(T^0)|0^0\rangle$$

In principle, D and $\mathbf{H}_1$ can be computed in terms of the known (unperturbed) cluster amplitudes and integrals involving the H_1 operator. The second-order equation that determines $T^{(2)}$ can also be expressed as a set of linear algebraic equations

$$\mathsf{D}\mathbf{t}^{(2)} = \mathbf{C} \tag{5.33}$$

where D was given above and $\mathbf{C}$ has elements

$$-\langle{}^{rs\cdots}_{\alpha\beta\cdots}|\exp(-T^0)\{[H_1, T^{(1)}] + \tfrac{1}{2}[[H, T^{(1)}], T^{(1)}]\}\exp(T^0)|0^0\rangle$$

Clearly, the evaluation of $\mathbf{C}$ requires that Eq. (5.32) first be solved for $T^{(1)}$. Then given $T^{(1)}$ and $T^{(2)}$, Eqs. (5.28) and (5.30) can be used to obtain the desired first- and second-order response properties as $E^{(1)}$ and $E^{(2)}$, respectively. We should point out that the term $T^{(2)}$ arising in this CC development has no analog in the MCSCF treatment given earlier. The absence of such quadratic terms in the MCSCF analog arises because, even if the energy expression given in Eq. (5.7) contained the term $-i\langle 0|[S^{(2)} + \lambda^{(2)}, H]|0\rangle$, it would vanish by the GBT. In the CC treatment of $E^{(2)}$ one needs both $T^{(1)}$ and $T^{(2)}$ because the CC wavefunction does not obey a GBT. We should also mention that, unlike the analogous result for the MCSCF response properties, the CC linear response energy $E^{(1)}$ is not simply equal to the average value of H_1. The term $\langle 0^0|\exp(-T^0)[H, T^{(1)}]\exp(T^0)|0^0\rangle$ has no counterpart in the MCSCF expression for $E^{(1)}$. In the event that the CC unperturbed energy $\langle 0^0|\exp(-T^0)H\exp(T^0)|0^0\rangle$ were stationary with respect to variations in T, this term would vanish.

F. PERTURBATIVE CALCULATION OF RESPONSE PROPERTIES

The RSPT or MBPT approach to computing response properties for atomic and molecular system is, in principle, straightforward (Kelly, 1969; Barlett and Silver, 1975). The perturbed Hamiltonian $H + \alpha H_1$ is decomposed into an unperturbed part H^0, which is most commonly taken to be a HF Hamiltonian, and a perturbation that contains both αH_1 and $(H - H^0)$:

$$H(\alpha) = H^0 + \alpha H_1 + H - H^0 \tag{5.34}$$

Then RSPT or MBPT is employed, as discussed in Chapter 3, to calculate perturbation corrections to the system energy. However, the terms $E_j^{(n)}$ are no longer simply grouped together according to their order in the total perturbation $\alpha H_1 + (H - H^0)$ but rather they are regrouped and labeled by two order indices $E_j^{(n,m)}$, which tell their separate orders in αH_1 and $(H - H^0)$, respectively. This additional decomposition is introduced because it is not practical to formulate a perturbation theory of the system's response to αH_1 in terms of the exact eigenstates of the full H.

The desired first- and second-order response properties of the state $|j\rangle$ are calculated by summing $E_j^{(1,m)}$ and $E_j^{(2,m)}$, respectively, over the index m (labeling order in $H - H^0$):

$$E^{(1)} = \sum_{m=0}^{\infty} E_j^{(1,m)} \tag{5.35}$$

$$E^{(2)} = \sum_{m=0}^{\infty} E_j^{(2,m)} \tag{5.36}$$

For practical reasons related to difficulty and expense in evaluating the higher-order contributions to $E_j^{(1,m)}$ and $E_j^{(2,m)}$, the index m is usually limited to rather small values.

Either the algebraic methods of RSPT or the diagrammatic methods of MBPT can be used to evaluate $E_j^{(1,m)}$ and $E_j^{(2,m)}$, as described in Chapter 3 in terms of the usual orbital energies, two-electron integrals, and one-electron integrals involving H_1 $(\langle\phi_r|H_1|\phi_s\rangle)$. Because both forms of perturbation theory yield energies that are size consistent, the evaluation of response properties as $E^{(1)}$ and $E^{(2)}$ guarantees that these properties will also be size consistent. As an example of how second-order properties may be evaluated, we display in Fig. 5.1 for a set of HF orbitals all of the zeroth- and first-order (in electron interaction) diagrams appropriate to a second-order response property whose perturbation operator [αH_1 of Eq. (5.2)] is denoted by a square figure. The evaluation of each of these diagrams is treated in the

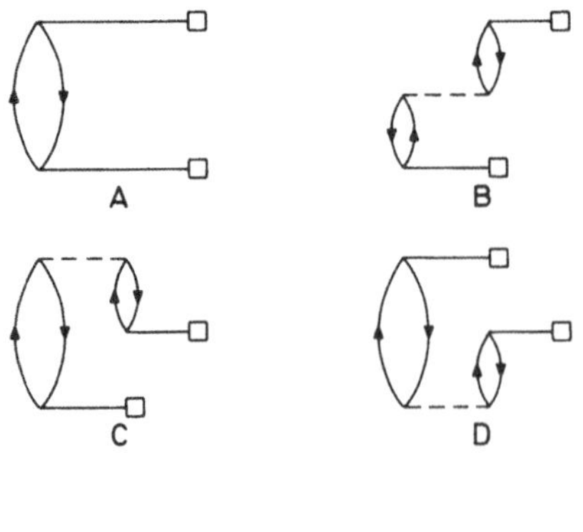

FIG. 5.1. All zeroth- and first-order diagrams for a second-order response property.

manner described in Table II of Chapter 3 with the matrix elements of the (one-electron) perturbation αH_1 being treated as the one-electron operator V. For example, the value of diagram D in Fig. 5.1 is given by

$$D = (-1)^{2+2} \sum_{\substack{pq \\ \gamma\beta}} \frac{\langle\beta|\alpha H_1|p\rangle\langle pq||\beta\gamma\rangle\langle\gamma|\alpha H_1|q\rangle}{(\varepsilon_\gamma + \varepsilon_\beta - \varepsilon_p - \varepsilon_q)(\varepsilon_\beta - \varepsilon_q)} \tag{5.37}$$

The evaluation of all the diagrams in Fig. 5.1 would thus give the desired second-order property consistent through first order in electronic interaction.

G. MOLECULAR GRADIENTS AND FORCE CONSTANTS

The determination of minima and saddle points on the potential-energy surface of a molecule plays an important role (Schaefer and Miller, 1977, Chapter 4) in describing the electronic structure and chemical reactivity of molecules. In this section, we show how such stationary points on a molecule's potential energy surface may be found by using an approach similar to that employed in Section 5.B. We first consider how the electronic Hamiltonian changes when the nuclear positions are changed from an initial set of positions, $\mathbf{R}_A^0$ to $\mathbf{R}_A$, i.e., $\mathbf{R}_A \to \mathbf{R}_A^0 + \mathbf{u}_A$. The electron–nuclear interaction is the only term in the Hamiltonian that depends explicitly on the nuclear position. Performing a Taylor expansion of this potential about the point $\mathbf{R}_A^0$, we obtain

$$\begin{aligned} |\mathbf{r} - \mathbf{R}_A|^{-1} = |\mathbf{r} - \mathbf{R}_A^0 - \mathbf{u}_A|^{-1} &= |\mathbf{r} - \mathbf{R}_A^0|^{-1} - (\mathbf{u}_A \cdot \nabla)|\mathbf{r} - \mathbf{R}_A^0|^{-1} \\ &+ \tfrac{1}{2}(\mathbf{u}_A \cdot \nabla)^2|\mathbf{r} - \mathbf{R}_A^0|^{-1} + O(\mathbf{u}_A^3) \end{aligned} \tag{5.38}$$

We may thus identify the changes in the electronic Hamiltonian through second order in the nuclear displacements ($\mathbf{u}_A$) as

$$V_1 = \sum_{\substack{A \\ ts}} + Z_A\langle\phi_t|[(\mathbf{u}_A \cdot \nabla)|\mathbf{r} - \mathbf{R}_A^0|^{-1}]|\phi_s\rangle t^+ s \tag{5.39}$$

$$V_2 = \sum_{\substack{A \\ ts}} - \tfrac{1}{2}Z_A\langle\phi_t|(\mathbf{u}_A \cdot \nabla)^2|\mathbf{r} - \mathbf{R}_A^0|^{-1}|\phi_s\rangle t^+ s \tag{5.40}$$

Here, V_1 clearly represents the forces on the electrons due to the nuclear displacement, whereas V_2 describes electric-field gradient terms induced by movement of the nuclei. A stationary point on the potential energy surface occurs when the average value of the first-order term in zero:

$$\langle 0|V_1|0\rangle = 0 \tag{5.41}$$

As demonstrated below, stationary points on the potential-energy surface may be determined in a quadratically convergent procedure using an analytical expression for the total energy that is accurate through second order in the nuclear displacement, and a Newton–Raphson procedure to determine the step length of the nuclear displacement. We now develop a procedure for carrying out such gradient calculations when $|0\rangle$ refers to a MCSCF wavefunction. Since the changes in the electronic Hamiltonian are determined in Eqs. (5.39) and (5.40) through second order in the nuclear displacement, an analytical expression of the total energy through second order in the nuclear displacement may be determined from the coupled multiconfiguration HF expression for the total energy given in Eq. (5.15) once αH_1 is identified as V_1 and $\alpha^2 H_2$ as V_2. The first-order term $\alpha\langle 0|H_1|0\rangle$ in Eq. (5.15) may be written as

$$\langle 0|V_1|0\rangle = \sum_A \mathbf{u}_A \cdot \mathbf{V}_{1A} \tag{5.42}$$

where the cartesian components of the force vector for displacement of nucleus A are

$$\mathbf{V}_{1A} = (V^x_{1A}, V^y_{1A}, V^z_{1A}) \tag{5.43}$$

with

$$V^i_{1A} = \sum_{ts} Z_A \langle \phi_t | \left[\frac{\partial}{\partial i} |\mathbf{r} - \mathbf{R}^0_A|^{-1} \right] |\phi_s\rangle \langle 0|t^+ s|0\rangle, \qquad i = x, y, z \tag{5.44}$$

The second-order term $\alpha^2\langle 0|H_2|0\rangle$ becomes

$$\langle 0|V_2|0\rangle = \sum_A \mathbf{u}_A \cdot \mathbf{V}_{2A} \cdot \mathbf{u}_A \tag{5.45}$$

where $\mathbf{V}_{2A}$ is a tensor operator, the components of which are defined as

$$V^{ij}_{2A} = \sum_{ts} -\tfrac{1}{2} Z_A \langle \phi_t | \left[\frac{\partial^2}{\partial i\, \partial j} |\mathbf{r} - \mathbf{R}^0_A|^{-1} \right] |\phi_s\rangle \langle 0|t^+ s|0\rangle, \qquad i, j = x, y, z \tag{5.46}$$

The matrix $\alpha\mathbf{F}$ given in Eq. (5.11) may similarly be written as

$$\alpha\mathbf{F} = \sum_A \mathbf{u}_A \cdot \mathbf{F}_A \tag{5.47}$$

where

$$\mathbf{F}^i_A = \sum_{ts} Z_A \langle \phi_t | \left[\frac{\partial}{\partial i} |\mathbf{r} - \mathbf{R}^0_A|^{-1} \right] |\phi_s\rangle \langle 0|[Q, t^+ s]|0\rangle, \qquad i = x, y, z \tag{5.48}$$

and an analogous expression can be written for the $\alpha\mathbf{G}$ matrix of Eq. (5.12). The last term of Eq. (5.15) therefore can be written as

$$\sum_{AB} \mathbf{u}_A \cdot \mathbf{D}_{AB} \mathbf{u}_B \tag{5.49}$$

where the tensor operator $\mathbf{D}_{AB}$ is defined through its components as

$$D^{ij}_{AB} = (\mathbf{F}^i_A \mathbf{G}^i_A)(\mathsf{A} - \mathsf{B})^{-1}\begin{pmatrix}\mathbf{F}^j_B \\ \mathbf{G}^j_B\end{pmatrix}, \qquad i, j = x, y, z \tag{5.50}$$

The total energy that contains all terms through second order in the nuclear displacement may thus be expressed as

$$E(\mathbf{u}) = \langle 0|H|0\rangle + \sum_A \mathbf{u}_A \cdot \mathbf{V}_{1A} + \sum_{AB} \mathbf{u}_A \cdot \mathbf{M}_{AB} \cdot \mathbf{u}_B \tag{5.51}$$

where

$$\mathbf{M}_{AB} = \mathbf{V}_{2A}\delta_{AB} - \mathbf{D}_{AB} \tag{5.52}$$

is the force constant matrix. This expression contains Hellmann–Feynman force terms in $\mathbf{V}_{1A}$, field gradient terms in $\mathbf{V}_{2A}$, as well as terms in $\mathbf{D}_{AB}$ that describe how the MCSCF orbitals and CI coefficients respond to displacements of the nuclei. A stationary point on the molecular potential energy surface is determined when $\delta E(\mathbf{u}) = 0$. Neglecting third- and higher-order terms in the energy function given in Eq. (5.51) and differentiating with respect to $\mathbf{u}$ thus gives

$$\mathbf{V}_1 + 2\mathsf{M}\mathbf{u} = 0 \tag{5.53}$$

where $\mathbf{V}_1$ and $\mathbf{u}$ are column vectors containing the elements V^x_{1A}, V^y_{1A}, V^z_{1A}, $V^x_{1B}, \ldots$ and u^x_A, u^y_A, u^z_A, $u^x_B, \ldots$, respectively. The elements of the matrix M are defined as the components of the tensor operator $\mathbf{M}_{AB}$ in Eq. (5.52):

$$(\mathsf{M})_{Ai,Bj} \equiv M^{ij}_{AB}, \qquad i, j = x, y, z \tag{5.54}$$

The nuclear displacements are thus given as

$$\mathbf{u} = -\tfrac{1}{2}\mathsf{M}^{-1}\mathbf{V}_1 \tag{5.55}$$

In the above derivation we have assumed that the atomic orbital basis employed in forming the MCSCF orbitals was complete. This assumption allowed us to write [in Eqs. (5.39) and (5.40)] the Hamiltonian both at $\mathbf{R}^0_A$ and $\mathbf{R}^0_A + \mathbf{u}_A$ in terms of the MCSCF orbitals, which were obtained from an MCSCF calculation performed at the "starting" geometry $\mathbf{R}^0_A$. In most molecular calculations, limited basis sets are used and the basis therefore depends on the nuclear positions. This dependence was not considered in the above derivation although it may be quite important depending on the basis set used in any particular calculation. Let us now assume that we shall attempt to describe the potential energy surface of a molecule by using an atomic orbital basis that is attached to the atomic nuclei and that thus moves with the nuclei. The above described formalism will be useful in locating the desired stationary points on the potential surface if both the first and second derivatives (with respect to nuclear displacement) of the dominant basis orbitals can be expanded in this basis. This can be seen by considering that the coulomb potential $|\mathbf{r} - \mathbf{R}_A|^{-1}$ of Eq. (5.38), when integrated over an electronic charge density $\rho(\mathbf{r})$, yields an interaction energy that can be expanded in powers of $\mathbf{u}_A \equiv \mathbf{R}_A - \mathbf{R}_A^0$ either by expanding $|\mathbf{r} - \mathbf{R}_A|^{-1}$ as in Eq. (5.38) or by expanding the charge density $\rho(\mathbf{r} - \mathbf{u}_A)$. The expansion of this charge density then gives rise to the derivatives of the atomic basis orbitals. This criterion—that the first and second derivatives of the important basis functions can be expanded in the same basis—may, of course, in principle never be met. For practical purposes we can, however, fulfill the criterion if the basis consists of a set of gaussian functions. We know that the nuclear displacement derivative of a gaussian function just is another gaussian with one higher angular momentum value. Thus by including such gaussian basis functions of higher angular momentum in the original basis, we could gaurantee that the derivatives of the important gaussian atomic orbitals will indeed by described within our finite basis. If the derivatives of the basis functions cannot be expanded in the basis, the fulfillment of Eq. (5.41) may not lead to an accurate stationary point. As an alternative to including in the atomic orbital basis sufficient flexibility to describe the first and second derivatives of the more important basis functions, one can explicitly evaluate derivatives of the one- and two-electron integrals (Thomsen and Swanstrøm, 1973). Suppose, for example, that *s*, *p*, and *d* atomic orbitals were used in a calculation on CH_2. Even if only the *s* and *p* orbitals were important in describing the orbitals having nonneglible occupation numbers, one would have to include full sets of *d* and *f* orbitals in the basis to guarantee that the second derivatives of the *s* and *p* functions could be described. As a result, many two-electron integrals involving *d* and *f* functions would have to be computed over the atomic orbital basis. On the other hand, this can be avoided by calculating only the first and second derivatives of the integrals over the *s* and *p* orbitals. These derivatives would then involve a very restricted subset of integrals containing *d* and *f* functions. For example, the second derivative of $\langle pp|pp\rangle$ would involve $\langle pp|pf\rangle$ and $\langle pp|dd\rangle$ integrals; integrals such as $\langle dd|dd\rangle$ or $\langle dd|fp\rangle$ or $\langle ff|ff\rangle$ could not arise. The smaller number of difficult integrals arising in approaches that explicitly evaluate integral derivatives rather than those using very large basis sets has made these integral derivative schemes more commonly used in state-of-the-art calculations.

PROBLEMS

5.1 Determine the excitation energies and transition moments for HeH^+ using the full CI calculation for HeH^+ again making use of SCF data of Problem 2.1. The nonvanishing matrix elements of the dipole operator

$\mathbf{r}(x, y, z)$ in the atomic basis are

$$\langle 1s_{\mathrm{H}}|z|1s_{\mathrm{He}}\rangle = 0.2854, \qquad \langle 1s_{\mathrm{H}}|z|1s_{\mathrm{H}}\rangle = 1.4$$

1. Determine the matrix elements of $\mathbf{r}$ in the SCF orbital basis.

In Problem 2.4, a full CI calculation was carried out on HeH^+ in the SCF orbital basis.

2. Use the results of Problem 2.4 to determine the excitation energies and transition moments from the ground state to the two excited singlet states of HeH^+.

3. Determine the frequency-independent polarizability and the frequency-dependent polarizability at a frequency $E = 0.1$ a.u. for HeH^+.

5.2 Carry out a coupled Hartree-Fock (CHF) calculation of the frequency-independent polarizability tensor for the closed-shell HeH^+ system. To achieve this goal, follow the steps given below.

1. Show that

$$\langle 0|[\mathbf{r}, m_\alpha^+ \alpha_\alpha + m_\beta^+ \alpha_\beta]10\rangle = 2\langle \phi_\alpha|\mathbf{r}|\phi_m\rangle$$

2. Use the A_{11} and B_{11} matrix elements derived in Problem 2.2 to show that the A_{11} and B_{11} matrix elements in a CHF calculation may be written as

$$\begin{aligned}(A_{11})_{n\beta,m\alpha} &= \langle 0|[\beta_\alpha^+ n_\alpha + \beta_\beta^+ n_\beta, H, m_\alpha^+ \alpha_\alpha + m_\beta^+ \alpha_\beta]|0\rangle \\ &= 2((\varepsilon_m - \varepsilon_\alpha)\delta_{mn}\delta_{\alpha\beta} + 2\langle n\alpha|\beta m\rangle - \langle n\alpha|m\beta\rangle)) \\ (B_{11})_{n\beta,m\alpha} &= \langle 0|[n_\alpha^+ \alpha_\alpha + n_\beta^+ \beta_\beta, H, m_\alpha^+ \alpha_\alpha + m_\beta^+ \alpha_\beta]|0\rangle \\ &= 2(\langle mn|\beta\alpha\rangle - 2\langle mn|\alpha\beta\rangle)\end{aligned}$$

To obtain these results you must use the definition of the Fock potential given in Eq. (2.92) and assume that the orbitals are HF orbitals.

Now carry out the CHF calculation on HeH^+, using the single zeta Slater basis and the SCF data given in Problem 2.1. The nonvanishing matrix elements of $\mathbf{r}(x, y, z)$ in the SCF basis are given in Solution 5.1, part 1.

3. Evaluate the matrix elements of $\mathbf{r}$, which enter into the CHF calculation.

4. Evaluate the A_{11} and B_{11} matrix elements.

5. Determine the frequency-independent polarizability tensor in the CHF approximation for HeH^+.

5.3 Carry out a CMCHF calculation of the frequency-independent polarizability tensor.

1. Show that

$$\langle 0|[\mathbf{r}, p_\alpha^+ q_\alpha + p_\beta^+ q_\beta]|0\rangle = \sum_s \{\mathbf{r}_{sp}\langle 0|s_\alpha^+ q_\alpha + s_\beta^+ q_\beta|0\rangle - \mathbf{r}_{qs}\langle 0|p_\alpha^+ s_\alpha + p_\beta^+ s_\beta|0\rangle\}$$

2. Show that

$$\langle 0|[\mathbf{r}, 1n\rangle\langle 0|]|0\rangle = \sum_{pq} \mathbf{r}_{pq}\langle 0|p_\alpha^+ q_\alpha + p_\beta^+ q_\beta|n\rangle$$

3. Indicate the simplifications that occur in the formulas for the A and B matrix elements in Eqs. (2.29) and (2.30), when converged MCSCF orbitals are used for evaluating the A and B matrices.

Now carry out the CMCHF calculation of the frequency-independent polarizability tensor for HeH^+, using the minimum basis given in Problem 2.1. The multiconfiguration reference state includes the two configurations $1\sigma^2$ and $2\sigma^2$. A MCSCF calculation using these two configurations was carried out in Problem 2.6. The one- and two-electron integrals in the MCSCF basis are given below:

$$\langle 1|h|1\rangle = -2.6119, \qquad \langle 2|h|2\rangle = -1.3193, \qquad \langle 1|h|2\rangle = 0.2078$$
$$\langle 11|11\rangle = 0.9521, \qquad \langle 12|12\rangle = 0.6100, \qquad \langle 12|11\rangle = -0.1963$$
$$\langle 11|22\rangle = 0.1298, \qquad \langle 22|21\rangle = -0.0069, \qquad \langle 22|22\rangle = 0.6161,$$

where 1 and 2 denote the 1σ and 2σ orbitals, respectively, which are

$$1\sigma = 0.8920\ 1s_{\rm He} + 0.1701\ 1s_{\rm H}, \qquad 2\sigma = -0.8410\ 1s_{\rm He} + 1.2140\ 1s_{\rm H}$$

The MCSCF states are

$$|0\rangle = 0.9984|1\sigma^2\rangle - 0.0574|2\sigma^2\rangle, \qquad E_0 = -2.8506$$
$$|1\rangle = 0.0574|1\sigma^2\rangle + 0.9984|2\sigma^2\rangle, \qquad E_1 = -0.5863$$

4. Calculate the nonvanishing one- and two-electron density and transition density matrix elements of the form

$$\langle i|r^+s^+tu|0\rangle, \qquad \langle i|r^+s|0\rangle, \langle 0|r^+s|i\rangle, \quad i = |0\rangle, |1\rangle$$

The matrix elements of $\mathbf{r}$ in the atomic basis are given in Problem 5.1.

5. Calculate the matrix elements of $\mathbf{r}$ in the MCSCF basis.

6. Calculate the numerical values of matrix elements given in questions 1 and 2.

7. Determine the A and B matrix elements.

8. Determine the frequency-independent polarizability tensor in the CMCHF approximation.

SOLUTIONS

5.1

1. $\langle 1|z|1\rangle = 2(0.9000)(0.1584)(0.2854) + (0.1584)^2(1.4) = 0.1165$

$\langle 2|z|2\rangle = 2(-0.8324)(1.2156)(0.2854) + (1.2156)^2(1.4) = 1.4911$

$\langle 1|z|2\rangle = [(0.9000)(1.2156) - (0.8324)(0.1584)]0.2854 + (0.1584)(1.2156)(1.4) = 0.5442$

2. Excitation energies are 4.2792 − 3.2567 = 1.0225 and 4.2792 − 1.9497 = 2.3295. Transition moments:

$$\langle 1\sigma^2|z|1\sigma^2\rangle = 2\langle 1|z|1\rangle, \qquad \langle 2\sigma^2|z|2\sigma^2\rangle = 2\langle 2|z|2\rangle, \qquad \langle 1\sigma^2|z|2\sigma^2\rangle = 0$$

$$\langle 1\sigma 2\sigma|z|1\sigma 2\sigma\rangle = \langle 1|z|1\rangle + \langle 2|z|2\rangle, \qquad \langle 1\sigma^2|z|1\sigma 2\sigma\rangle = \frac{2}{\sqrt{2}}\langle 1|z|2\rangle,$$

$$\langle 2\sigma^2|z|1\sigma 2\sigma\rangle = \frac{2}{\sqrt{2}}\langle 2|z|1\rangle.$$

Therefore

$$\begin{aligned}\langle 0|z|1\rangle = {} & (0.9982)(-0.0261)2(0.1165) + (-0.0573)(-0.2098)2(1.4911) \\ & + (0.0143)(0.9772)(0.1165 + 1.4911) + [(0.9982)(0.9772) \\ & + (0.0143)(-0.0261) + (-0.0573)(0.9772) \\ & + (-0.2098)(0.0143)]\left(\frac{0.5442}{\sqrt{2}}\right)2 \\ = {} & 0.7578\end{aligned}$$

$$\begin{aligned}\langle 0|z|2\rangle = {} & (0.9982)(0.0530)2(0.1165) + (-0.0573)(0.9761)2(1.4911) \\ & - (0.0143)(0.2109)(0.1165 + 1.4911) + [(0.9982)(0.2109) \\ & + (0.0143)(0.0530) + (-.0573)(0.2109) \\ & + (0.0143)(0.9761)]\left(\frac{0.5442}{\sqrt{2}}\right)2 \\ = {} & 0.0144\end{aligned}$$

3. $$\alpha_{zz}(E) = 2\sum_{n=1,2}\frac{|\langle 0|z|n\rangle|^2(E_n - E_0)}{(E_n - E_0)^2 - E}$$

$$E = 0, \qquad \alpha_{zz} = 2\left[\frac{0.5742}{1.0225} + \frac{0.0002}{2.3295}\right] = 1.1233$$

$$E = 0.1, \qquad \alpha_{zz} = 2\left[\frac{(0.5742)(1.0225)}{1.0456 - 0.01} + \frac{(.0002)(2.3295)}{5.4266 - 0.01}\right] = 1.1342$$

5.2

1. $$\begin{aligned}\langle 0|[\mathbf{r}, m_\alpha^+\alpha_\alpha + m_\beta^+\alpha_\beta]|0\rangle &= \sum_{pq}\langle\phi_p|\mathbf{r}|\phi_q\rangle\langle 0|[p^+q, m_\alpha^+\alpha_\alpha + m_\beta^+\alpha_\beta]|0\rangle \\ &= \sum_{pq}\langle\phi_p|\mathbf{r}|\phi_q\rangle\langle 0|p^+\alpha_\alpha\delta_{m_\alpha q} - m_\alpha^+ q\delta_{p\alpha_\alpha} \\ &\quad + p^+\alpha_\beta\delta_{qm_\beta} - m_\beta^+ q\delta_{p\alpha_\beta}|0\rangle \\ &= 2\langle\phi_\alpha|\mathbf{r}|\phi_m\rangle\end{aligned}$$

2. The definition of the Fock potential in Eq. (2.92) results in the following definition of the orbital energies

$$h_{rs} + \sum_{\gamma}(2\langle r\gamma|s\gamma\rangle - \langle r\gamma|\gamma s\rangle) = \delta_{rs}\varepsilon_r$$

and the A_{11} matrix element in Problem 2.2 therefore may be rewritten as

$$(A_{11})_{n\beta,m\alpha} = 2[(-\varepsilon_\alpha + \varepsilon_m)\delta_{\alpha\beta}\delta_{mn} + 2\langle n\alpha|\beta m\rangle - \langle n\alpha|m\beta\rangle]$$

3. Let 1 and 2 denote 1σ and 2σ, respectively. As was shown in Problem 5.1 $\langle 1|z|2\rangle = 0.5442$. *Hence*

$$\langle 0|z(2_\alpha^+ 1_\alpha + 2_\beta^+ 1_\beta)|0\rangle = 2(0.5442)$$

4. $(A_{11})_{21,21} = 2(\varepsilon_2 - \varepsilon_1 + 2\langle 21|12\rangle - \langle 21|21\rangle) = 2.1464,$
$(B_{11})_{21,21} = 2(\langle 22|11\rangle - 2\langle 22|11\rangle) = -0.2522$

5. Only the zz component of the polarizability tensor is nonvanishing. This component becomes $2 \cdot 4 \cdot 0.5442^2/(2.1464 + 0.2522) = 0.9878$.

5.3

1. Using $\mathbf{r} = \sum_{st} \mathbf{r}_{st}(s_\alpha^+ t_\alpha + s_\beta^+ t_\beta)$ we obtain

$$\langle 0|[\mathbf{r}, p_\alpha^+ q_\alpha + p_\beta^+ q_\beta]|0\rangle = \sum_{st} \mathbf{r}_{ts}\langle 0|[t_\alpha^+ s_\alpha + t_\beta^+ s_\beta, p_\alpha^+ q_\alpha + p_\beta^+ q_\beta]|0\rangle$$

Performing the commutations then leads immediately to the result asked for.

2. $\langle 0|[\mathbf{r}, |n\rangle\langle 0|]|0\rangle = \langle 0|\mathbf{r}|n\rangle = \sum_{pq} \mathbf{r}_{pq}\langle 0|p_\alpha^+ q_\alpha + p_\beta^+ q_\beta|n\rangle$

3. The A_{11} and B_{11} matrices in Eqs. (2.29) and (2.30) can, when convergence is reached and the GBT is obeyed, be evaluated directly using Eq. (2.42), which does contain the double commutator form.

4.

	$\langle i\| = \langle 0\|$	$\langle i\| = \langle 1\|$
$\langle i\|1_\alpha^+ 1_\alpha\|0\rangle$	0.9968	0.0573
$\langle i\|2_\alpha^+ 2_\alpha\|0\rangle$	0.0033	−0.0573
$\langle i\|1_\alpha^+ 1_\beta^+ 1_\beta 1_\alpha\|0\rangle$	0.9968	0.0573
$\langle i\|1_\alpha^+ 1_\beta^+ 2_\beta 2_\alpha\|0\rangle$	−0.0573	−0.0033
$\langle i\|2_\alpha^+ 2_\beta^+ 1_\beta 1_\alpha\|0\rangle$	−0.0573	0.9968
$\langle i\|2_\alpha^+ 2_\beta^+ 2_\beta 2_\alpha\|0\rangle$	0.0033	−0.0573

We also have $\langle 0|1_\alpha^+ 1_\alpha|1\rangle = 0.0573$, $\langle 0|2_\alpha^+ 2_\alpha|1\rangle = -0.0573$. See text below Solution 2.6, question 1.

5. $z_{11} = 0.1271$, $z_{22} = 1.4805$, $z_{12} = 0.5574$.

6. $\langle 0|[z, 2_\alpha^+ 1_\alpha + 2_\beta^+ 1_\beta]|0\rangle = 2z_{12}\{\langle 0|1_\alpha^+ 1_\alpha - 2_\alpha^+ 2_\alpha|0\rangle\} = 1.1076$
$\langle 0|[z, |1\rangle\langle 0|]|0\rangle = 2z_{11}\langle 0|1_\alpha^+ 1_\alpha|1\rangle + 2z_{22}\langle 0|2_\alpha^+ 2_\alpha|1\rangle = -0.1551$

7. Formulas for the A and B matrix element are derived in Solution 2.6, question 2:

$$\mathsf{A} = \begin{pmatrix} 2.1756 & 0.4018 \\ 0.4018 & 2.2643 \end{pmatrix}, \qquad \mathsf{B} = \begin{pmatrix} -0.2495 & -0.0230 \\ -0.0230 & 0 \end{pmatrix}$$

8. $$(\mathsf{A} - \mathsf{B})^{-1} = \begin{pmatrix} 0.4264 & -0.0800 \\ -0.0800 & 0.4566 \end{pmatrix}$$

The zz component of the frequency-independent polarizability becomes

$$\alpha_{zz} = 2(1.1076, -0.1551)\begin{pmatrix} 0.4264 & -0.0800 \\ -0.0800 & 0.4566 \end{pmatrix}\begin{pmatrix} 1.1076 \\ -0.1551 \end{pmatrix} = 1.1232$$

References

Bartlett, R. J., and G. D. Purvis, (1979). *Phys. Rev.* **A 20**, 1313.

Bartlett, R. J., and Silver, D. M. (1975). *Int. J. Quantum Chem.* **S9**, 183.

Dalgaard, E., and Jørgensen, P. (1978). *J. Chem. Phys.* **69**, 3833.

Kelly, H. P. (1969). *Adv. Chem. Phys.* **14**, 129.

Monkhorst, H. J. (1977). *Int. J. Quantum Chem.* **S11**, 421.

Schaefer, H. F., and Miller, W. H. eds. (1975). "Modern Theoretical Chemistry," Vol. 4. Plenum, New York.

Thomsen, K., and Swanstrøm, P. (1973). *Mol. Phys.* **26**, 735.

Chapter 6 | Green's Functions

A. INTRODUCTION

Having now seen how methods that are based upon stationary-state N-electron wavefunctions can be used to compute state energies and other physical properties, we turn to examine a class of so-called response functions or Green's functions (GFs) (Linderberg and Öhrn, 1973), which permit a direct calculation of transition properties. For example, the one-particle GF (electron propagator) yields ionization potentials and electron affinities, whereas the two particle GF (polarization propagator) provides us information about electronic excitation energies and oscillator strengths, which then can be used to calculate many other observables (e.g., polarizabilities and spin–spin coupling constants). The general definition of a GF belonging to the reference state $|0\rangle$ is given as

$$\langle\langle A(t); B\rangle\rangle = \mp\frac{i}{\hbar}\,\theta(t)\langle 0|A(t)B|0\rangle + \frac{i}{\hbar}\,\theta(-t)\langle 0|BA(t)|0\rangle \tag{6.1}$$

where $\theta(t)$ is the Heaviside step function

$$\theta(t) = \{1, t > 0; 0, t < 0\} \tag{6.2}$$

and A and B are arbitrary operators in the second quantization form. $A(t)$ is the Heisenberg representation of A,

$$A(t) = \exp(iHt/\hbar)A\exp(-iHt/\hbar) \tag{6.3}$$

and B is the Heisenberg operator at $t = 0$. If the operators A and B contain an even number of creation or annihilation operators (e.g., r^+s, or r^+t^+su) the plus sign is used in Eq. (6.1). For operators A and B having an odd number of such operators (e.g., $A = r^+$ or r^+t^+u, and $B = r$ or rut^+) the minus sign is used. The reasons for these choices are made clear below.

To understand the physical content of such GFs, we introduce between the $A(t)$ and B operators in Eq. (6.1) a resolution of the identity involving a complete set of eigenstates $|n\rangle$. By inserting these resolutions and assuming that the states $|n\rangle$ are eigenfunctions of H, we obtain

$$\langle\langle A(t); B\rangle\rangle = \mp\frac{i}{\hbar}\,\theta(t)\sum_n \langle 0|A|n\rangle\langle n|B|0\rangle \exp\left[i\frac{t}{\hbar}(E_0 - E_n)\right] + \frac{i}{\hbar}\,\theta(-t)\sum_n \langle 0|B|n\rangle\langle n|A|0\rangle \exp\left[i\frac{t}{\hbar}(E_n - E_0)\right] \quad (6.4)$$

The Fourier transform of $\langle\langle A(t)B\rangle\rangle$, is given as

$$\langle\langle A; B\rangle\rangle_E = \int_{-\infty}^{\infty} dt\, \langle\langle A(t); B\rangle\rangle \exp(iEt/\hbar) \quad (6.5)$$

If we straightforwardly insert Eq. (6.4) into Eq. (6.5) we encounter improper integrals. For example, the first term of Eq. (6.4) gives an improper integral of the type

$$\int_0^{\infty} dt \exp\left[i\frac{t}{\hbar}(E_0 - E_n + E)\right]$$

To overcome this problem we may define the Fourier transform to include a convergence factor $\exp(-|t|\eta)$, where η is a small, real, positive quantity. After the integration is performed, we can then take the limit $\eta \to 0_+$ (Mattuck, 1967). The Fourier transform of the GF may then be expressed as

$$\langle\langle A; B\rangle\rangle_E = \lim_{\eta\to 0_+} \sum_n \frac{\pm\langle 0|A|n\rangle\langle n|B|0\rangle}{E_0 - E_n + E + i\eta} + \sum_n \frac{\langle 0|B|n\rangle\langle n|A|0\rangle}{E_n - E_0 + E - i\eta} \quad (6.6)$$

A physical interpretation of the GF may now be obtained by considering the content of Eq. (6.6). If A and B are number conserving (i.e., they both contain equal numbers of creation and annihilation operators) then the states $|n\rangle$ must contain the same number of electrons as the reference state $|0\rangle$ to give a nonvanishing GF. However, if A contains, for example, one more creation operator than annihilation operator, then $|n\rangle$ must contain $N \pm 1$ electrons (notice that the fact that the second-quantized H is independent of N is now becoming convenient). From the frequency spectrum of $\langle\langle A; B\rangle\rangle_E$ it is clear that the GF contains information about energy differences. If A and B are of the one-particle excitation form r^+s, then poles of Eq. (6.6) occur at the energy differences $E_n - E_0$ referring to electronic excitation energies. The residues give the overlap amplitudes $\langle 0|A|n\rangle\langle n|B|0\rangle$, which, for example, express the electric dipole transition probabilities when A and B refer to the electronic dipole moment operator. If A is of the form r^+ (so that B is an annihilation operator s), then the energy differences arising in Eq. (6.6) fall into two classes. The first factor, which has $\langle 0|r^+|n\rangle\langle n|s|0\rangle$ as its amplitude, clearly has to do with ionization potentials $E_0^N - E_n^{N-1}$. The second factor, involving amplitudes $\langle 0|s|n\rangle\langle n|r^+|0\rangle$, relates to electron affinities $E_n^{N+1} - E_0^N$.

The time derivative of Eq. (6.1) may be written as

$$i\hbar \frac{d}{dt} \langle\langle A(t); B\rangle\rangle = \delta(t)\langle 0|\pm A(t)B + BA(t)|0\rangle + \langle\langle [A(t), H]B\rangle\rangle$$

$$= \delta(t)\langle 0|BA \pm AB|0\rangle + \langle\langle [A(t), H]B\rangle\rangle \tag{6.7}$$

where we have used the facts that $A(t)$ satisfies the Heisenberg equation of motion

$$i\hbar \frac{d}{dt} A(t) = [A(t), H] \tag{6.8}$$

and that the Heaviside function is the integral of the Dirac δ-function

$$\theta(t) = \int_{-\infty}^{t} \delta(\tau)\, d\tau \tag{6.9}$$

The Fourier transform of Eq. (6.7) then becomes [the definition of the Fourier transform of the GF always contains the $\exp(-\eta|t|)$ convergence factor, although henceforth we do not explicitly express this fact]

$$E\langle\langle A; B\rangle\rangle_E = \langle 0|BA \pm AB|0\rangle + \langle\langle [A, H]; B\rangle\rangle_E \tag{6.10}$$

As we see later, this result will prove useful in interrelating GFs when A and B refer to the position and momentum operator, respectively.

Although the above spectral representation of $\langle\langle A; B\rangle\rangle_E$ in Eq. (6.6) displays the content of its frequency dependence and amplitudes, this equation is not actually used to compute $\langle\langle A; B\rangle\rangle_E$. To do so would involve computing, by stationary-state methods described in earlier chapters, the energies and wavefunctions $|0\rangle$, E_0, $|n\rangle$, and E_n. The philosophy of the GF method is to avoid doing all of these state calculations by obtaining an equation that can be solved directly for $\langle\langle A; B\rangle\rangle_E$. In this manner one then attempts to obtain an object ($\langle\langle A; B\rangle\rangle_E$) that contains (through its poles and residues) state difference information directly.

B. SUPEROPERATOR ALGEBRA

1. Superoperator Resolvent

To demonstrate how one goes about finding an equation that permits $\langle\langle A; B\rangle\rangle_E$ to be directly computed, let us return to Eq. (6.1) and rewrite the time dependence of $A(t)$ as

$$\begin{aligned} A(t) &\equiv \exp\left(\frac{it}{\hbar}H\right)A\exp\left(-\frac{it}{\hbar}H\right) \\ &= A + \frac{it}{\hbar}[H,A] + \frac{1}{2!}\left(\frac{it}{\hbar}\right)^2[H,[H,A]] \\ &\quad + \frac{1}{3!}\left(\frac{it}{\hbar}\right)^3[H,[H,[H,A]]] + \cdots \\ &\equiv \exp\left(\frac{it}{\hbar}\hat{H}\right)A \end{aligned} \tag{6.11}$$

where the so-called superoperator $\hat{H}$ (Pickup and Goscinski, 1973) is defined by

$$\hat{H}A \equiv [H,A] \tag{6.12}$$

In terms of this superoperator, $\langle\langle A(t);B\rangle\rangle$ can be expressed as

$$\begin{aligned} \langle\langle A(t);B\rangle\rangle = &\mp\frac{i}{\hbar}\theta(t)\langle 0|\left(\exp\left(\frac{i}{\hbar}t\hat{H}\right)A\right)B|0\rangle \\ &+\frac{i}{\hbar}\theta(-t)\langle 0|B\exp\left(\frac{i}{\hbar}t\hat{H}\right)A|0\rangle \end{aligned} \tag{6.13}$$

[notice that the extra parentheses are needed in the first term on the right-hand side of Eq. (6.13) to ensure that $\hat{H}$ only operates on A]. The Fourier transform can now be carried out to yield

$$\langle\langle A;B\rangle\rangle_E = \pm\langle 0|((E\hat{1}+\hat{H})^{-1}A)B|0\rangle + \langle 0|B(E\hat{1}+\hat{H})^{-1}A|0\rangle \tag{6.14}$$

where the unit superoperator is defined by

$$\hat{1}A \equiv A \tag{6.15}$$

It is conventional to combine the two terms present on the right-hand side of Eq. (6.14) into a single factor by introducing the so-called superoperator binary product. This product, between two operators C and D, is defined as

$$(C|D) \equiv \langle 0|C^+D|0\rangle \pm \langle 0|DC^+|0\rangle \tag{6.16}$$

with the plus sign pertaining to cases when C and D contain odd numbers of creation or annihilation operators (e.g., r^+s^+t or u). With this definition, the above GF can be written as

$$\langle\langle A;B\rangle\rangle_E = (B^+|(E\hat{1}+\hat{H})^{-1}|A) \tag{6.17}$$

In writing $\langle\langle A; B\rangle\rangle_E$ in this way, we say that we have expressed the GF as a superoperator matrix element of the superoperator resolvent $(E\hat{1} + \hat{H})^{-1}$.

2. Complete Sets of Operators

The tools needed for evaluating the above matrix elements of the superoperator resolvent are based upon the idea of operators (of the same "type" as A and B^+) forming complete sets (Manne, 1977; Dalgaard, 1979). For example, if A and B^+ are number-conserving operators (e.g., r^+s), then the set of operators ($\alpha > \beta > \gamma > \cdots; p > q > r > \cdots$)

$$\{\mathbf{h}\} = \{1, p^+\alpha, p^+q^+\beta\alpha, p^+q^+r^+\gamma\beta\alpha, \ldots\} \tag{6.18}$$

when operating on an N-electron ket corresponding to a single determinant in which $\phi_\alpha, \phi_\beta, \phi_\gamma, \ldots$ are "occupied" and $\phi_p, \phi_q, \phi_r, \ldots$ are not occupied, forms a complete set of N-electron kets. Similarly ($\alpha > \beta > \gamma > \cdots; p > q > r > \cdots$)

$$\{\mathbf{h}\} = \{r^+, r^+p^+\alpha, r^+p^+q^+\alpha\beta, \ldots\} \tag{6.19}$$

and

$$\{\mathbf{h}\} = \{\alpha, \alpha\beta r^+, \alpha\beta\gamma r^+s^+ \ldots\} \tag{6.20}$$

form, respectively, complete sets of $(N \pm 1)$-electron kets when operating on the above "reference ket". Manne and Dalgaard have shown that the above sets of operators form complete sets of N- and $(N \pm 1)$-electron kets even when operating on a multiconfigurational reference state $|0\rangle$ as long as the reference ket (which defines $\alpha, \beta, \gamma, \ldots; p, q, r, \ldots$) is not orthogonal to $|0\rangle$.

The above results having to do with completeness of operator manifolds permit us to write a resolution of the identity as

$$1 = \sum_{kl} h_k|0\rangle\langle 0|\mathbf{h}^+\mathbf{h}|0\rangle_{kl}^{-1}\langle 0|h_l^+ \tag{6.21}$$

where the set $\{h_k\}$ is any of the above three sets of operator manifolds and $\langle 0|\mathbf{h}^+\mathbf{h}|0\rangle_{kl}^{-1}$ is the k, l element of the inverse of the matrix having elements $\langle 0|h_k^+h_l|0\rangle$. The completeness relation mentioned above cannot be used in a straightforward manner in manipulations having to do with the superoperator resolvent because the superoperator binary product appearing in $\langle\langle A; B\rangle\rangle_E$ is more complicated than the scalar product occurring in Eq. (6.21). The complete set of operators for N- and $(N \pm 1)$-electron kets may, however, be used to generate a resolution of the identity that can be used within the superoperator binary product. The completeness relation for a superoperator binary product may be written as

$$\hat{1} = |\mathbf{T}^+)(\mathbf{T}^+|\mathbf{T}^+)^{-1}(\mathbf{T}^+| = \sum_{kl}|T_k^+)(\mathbf{T}^+|\mathbf{T}^+)_{kl}^{-1}(T_l^+| \tag{6.22}$$

where $\mathbf{T}^+$ for one-electron creation or annihilation operators A and B^+ becomes ($\alpha > \beta > \gamma > \cdots ; p > q > r > \cdots$) (Dalgaard, 1979)

$$\begin{aligned}\{\mathbf{T}^+\} &= \{\mathbf{T}_1^+ ; \mathbf{T}_3^+ ; \mathbf{T}_5^+ ; \ldots\} \\ &= \{p^+, \alpha^+ ; p^+q^+\alpha, \alpha^+\beta^+p ; p^+q^+r^+\alpha\beta, \alpha^+\beta^+\gamma^+pq ; \ldots\} \end{aligned} \tag{6.23}$$

For number-conserving operators A and B^+, $\{\mathbf{T}^+\}$ becomes

$$\{\mathbf{T}^+\} = \{\mathbf{T}_2^+ ; \mathbf{T}_4^+ ; \ldots\} = \{p^+\alpha, \alpha^+p ; p^+q^+\alpha\beta ; \alpha^+\beta^+pq, \ldots\} \tag{6.24}$$

To better appreciate the meaning of Eq. (6.22), we write in detail some elements of the "overlap" matrix $(T_k^+ \mid T_l^+)$ for the one-electron addition operator case (recall the definition of the "occupied" and "unoccupied" orbitals, ϕ_α, ϕ_r)

$$(\alpha^+ \mid p^+) = \langle 0|\alpha p^+ + p^+\alpha|0\rangle = \delta_{\alpha p} = 0 \tag{6.25}$$

$$\begin{aligned}(\alpha^+ \mid p^+q^+\beta) &= \langle 0|\alpha p^+q^+\beta + p^+q^+\beta\alpha|0\rangle \\ &= \langle 0|p^+q^+\alpha\beta + p^+q^+\beta\alpha|0\rangle = 0\end{aligned} \tag{6.26}$$

$$\begin{aligned}(r^+ \mid p^+q^+\alpha) &= \langle 0|rp^+q^+\alpha + p^+q^+\alpha r|0\rangle \\ &= \delta_{rp}\langle 0|q^+\alpha|0\rangle - \delta_{rq}\langle 0|p^+\alpha|0\rangle.\end{aligned} \tag{6.27}$$

It is clear from the above equations that, in the superoperator binary product, each of the operators contributes both to the $(N + 1)$- and to the $(N - 1)$-electron aspects of the problem. For example, in writing the binary product $(r^+ \mid p^+q^+\alpha)$, we find $\langle 0| r$, which refers to the adjoint of an $(N + 1)$-electron ket, while $r|0\rangle$ becomes an $(N - 1)$-electron ket.

3. The Superoperator Resolvent

In summary, the idea of a complete set of operators has been extended to the superoperator binary product so as to introduce the powerful concept of a completeness relation. This completeness relation can now be exploited to derive an equation that permits $\langle\langle A; B\rangle\rangle_E$ to be expressed in a computationally more useful form (Simons, 1976). We begin by writing the identity

$$(\mathbf{T}^+ \mid \mathbf{T}^+) = (\mathbf{T}^+ |(E\hat{1} + \hat{H})(E\hat{1} + \hat{H})^{-1}| \mathbf{T}^+) \tag{6.28}$$

which, by inserting the resolution of the identity in Eq. (6.22), becomes

$$(\mathbf{T}^+ \mid \mathbf{T}^+) = (\mathbf{T}^+ | E\hat{1} + \hat{H} | \mathbf{T}^+)(\mathbf{T}^+ \mid \mathbf{T}^+)^{-1}(\mathbf{T}^+ |(E\hat{1} + \hat{H})^{-1}| \mathbf{T}^+) \tag{6.29}$$

This equation can be arranged to yield

$$(\mathbf{T}^+ |(E\hat{1} + \hat{H})^{-1}| \mathbf{T}^+) = (\mathbf{T}^+ \mid \mathbf{T}^+)(\mathbf{T}^+ | E\hat{1} + \hat{H} | \mathbf{T}^+)^{-1}(\mathbf{T}^+ \mid \mathbf{T}^+) \tag{6.30}$$

The operators B^+ and A, which define the desired GF in Eq. (6.17), when operating on $|0\rangle$, can be expanded in terms of the set $\{\mathbf{T}^+\}$ operating on $|0\rangle$:

$$A|0\rangle = \mathbf{T}^+|0\rangle(\mathbf{T}^+|\mathbf{T}^+)^{-1}(\mathbf{T}^+|A) \tag{6.31}$$

and the GF in Eq. (6.17) may thus, using Eqs. (6.30) and (6.31), be written as

$$\begin{aligned}\langle\langle A;B\rangle\rangle_E &= (B^+|(E\hat{I} + \hat{H})^{-1}|A) \\ &= (B^+|\mathbf{T}^+)(\mathbf{T}^+|\mathbf{T}^+)^{-1}(\mathbf{T}^+|(E\hat{I} - \hat{H})^{-1}|\mathbf{T}^+)(\mathbf{T}^+|\mathbf{T}^+)^{-1}(\mathbf{T}+|A) \\ &= (B^+|\mathbf{T}^+)(\mathbf{T}^+|E\hat{I} + \hat{H}|\mathbf{T}^+)^{-1}(\mathbf{T}^+|A)\end{aligned} \tag{6.32}$$

Equation (6.32) constitutes the working equation for deriving approximate forms for the GF. Notice that the original GF, which involved the matrix representative of an inverse superoperator $(E\hat{I} + \hat{H})^{-1}$, has been expressed in terms of the elements $(B^+|T_k^+)$, $(T_l^+|A)$, the "overlap" $(T_k^+|T_l^+)$, and matrix elements of the superoperator Hamiltonian $(T_k^{\dagger}|\hat{H}|T_l^+)$. These latter two matrices are analogous to the expressions that give ordinary resolvent matrix elements in terms of configuration interaction Hamiltonian matrix elements and configuration overlaps.

4. Pole and Residue Analysis

From Eq. (6.32), which expresses the desired GF, it is clear that the pole structure (values of E at which $\langle\langle A;B\rangle\rangle_E$ has poles) is determined entirely by the matrix $(\mathbf{T}^+|E\hat{I} + \hat{H}|\mathbf{T}^+)^{-1}$. This matrix has poles when $\det[(\mathbf{T}^+|E\hat{I} + \hat{H}|\mathbf{T}^+)]$ vanishes. Thus, the problem of finding the poles of $\langle\langle A;B\rangle\rangle_E$, which give ionization or excitation energies, can be solved by examining the superoperator generalized eigenvalue problem

$$\sum_l (T_k^+|\hat{H}|T_l^+)U_{lj} = -E_j \sum_l (T_k^+|T_l^+)U_{lj} \tag{6.33}$$

which in matrix notation may be written as

$$\hat{\mathsf{H}}\mathbf{U}_j = -E_j\mathsf{S}\mathbf{U}_j \tag{6.34}$$

The poles of $\langle\langle A;B\rangle\rangle_E$ occur thus at the eigenvalues $E = E_j$ of Eq. (6.33) and the eigenvectors enter in the evaluation of the corresponding residues. To illustrate how the residues can be determined, we rewrite the GF in Eq. (6.32) so as to be in spectral form, assuming that $\hat{\mathsf{H}}$ is hermitian and that S is positive definite (this is not always the case as we discuss in Section 6.E.2.a). Premultiplying Eq. (6.34) by $\mathsf{S}^{-1/2}$ gives

$$(\mathsf{S}^{-1/2}\hat{\mathsf{H}}\mathsf{S}^{-1/2})(\mathsf{S}^{1/2}\mathbf{U}_j) = -E_j(\mathsf{S}^{1/2}\mathbf{U}_j) \tag{6.35}$$

The vectors $\mathbf{V}_j = \mathsf{S}^{1/2}U_j$ are ordinary eigenvectors of $\mathsf{H}' \equiv \mathsf{S}^{-1/2}\hat{\mathsf{H}}\mathsf{S}^{-1/2}$, and therefore H' can be expressed in spectral form as

$$\hat{\mathsf{H}}' = \sum_j \mathbf{V}_j(-E_j)\mathbf{V}_j^+ \tag{6.36}$$

Because the $\mathbf{V}_j$ are eigenvectors of a hermitian matrix (H'), they form a unitary matrix (V), which diagonalizes H'. Thus, one can write the resolvent matrix as

$$\begin{aligned}(E\mathsf{S} + \hat{\mathsf{H}})^{-1} &= \mathsf{S}^{-1/2}(E\hat{1} + \hat{\mathsf{H}}')^{-1}\mathsf{S}^{-1/2} = \mathsf{S}^{-1/2}\mathsf{V}(E\hat{1} - \mathsf{E})^{-1}\mathsf{V}^+S^{-1/2} \\ &= \mathsf{U}(E\hat{1} - \mathsf{E})^{-1}\mathsf{U}^+ \end{aligned} \tag{6.37}$$

where the diagonal matrix E contains the eigenvalues E_j.

By using Eq. (6.37), the expression for $\langle\langle A;B\rangle\rangle_E$ given in Eq. (6.32) can be rewritten in a form that clearly displays its pole and residue structure:

$$\langle\langle A;B\rangle\rangle_E = (B^+|\mathbf{T}^+)\mathsf{U}(E\hat{1} - \mathsf{E})^{-1}\mathsf{U}^+(\mathbf{T}^+|A) \tag{6.38}$$

Thus, the residue at pole E_j is given by

$$\sum_{k,l} (B^+|T_k^+)U_{kj}U_{jl}^+(T_l^+|A) \tag{6.39}$$

C. APPROXIMATION METHODS

1. Operator Manifold Truncation

Although the above equations, in principle, permit one to find the poles and residues of any GF (defined by the choice of B^+ and A), it is never really possible to employ a complete set of operators $\{\mathbf{T}_k^+\}$. Therefore, one is faced both with making some physically motivated choice of a finite number of such $\{\mathbf{T}_k^+\}$ operators and with choosing a reasonably accurate reference wavefunction $|0\rangle$. Clearly the choice of $|0\rangle$ dictates which excitation or ionization energies one obtains from the poles of $\langle\langle A;B\rangle\rangle_E$. The choice of B^+ and A determine whether one is interested in single-particle excited states ($A = i^+j$), primary ionization potentials ($A = j$), or shake-up ionization potentials ($A = ijk^+$). For example, by using as $|0\rangle$ the $2s^2$ configuration and $A = j^+$, one can obtain ionization energies to the $2s^22p$ or $2s^1$ and other anion and cation states; with $A = m^+\alpha$, the $2snp$ excited states may be reached. The truncation of the complete operator set $\{\mathbf{T}_k^+\}$ then determines, together with the approximation made to get $|0\rangle$, the accuracy to which the resultant poles of $\langle\langle A;B\rangle\rangle_E$ describe the excitation or ionization energies and their corresponding residues. Choices of $\{\mathbf{T}^+\}$ must, of course, take into consideration the space and spin symmetry of the states generated by $\mathbf{T}^+|0\rangle$. As a result, the inherent symmetry of each

$\mathbf{T}^+$ must be coupled to that of $|0\rangle$ to give pure symmetry excited or ionic states. Because the reasons for making specific choices of A and B^+ are rather clear, we now focus on explaining the strategies for choosing $|0\rangle$ and truncations of $\{\mathbf{T}_k^+\}$. In the following sections and in our treatment of the polarization propagator, we consider two different approaches for attacking this problem. The first is based on a perturbation analysis while the second is based on selecting a multiconfiguration reference state and an appropriate projection manifold.

2. Order Analysis

The most widely used, and historically older, approach involves perturbation analysis of the GF using RSPT to obtain elements of $(\mathbf{T}^+|E\hat{1} + \hat{H}|\mathbf{T}^+)$ and $(B^+|\mathbf{T}^+)(\mathbf{T}^+|A)$ correct through a chosen order (order is then assumed to be related to accuracy). By decomposing the electronic Hamiltonian H and the reference wavefunction $|0\rangle$ in perturbation series

$$H = H^0 + V \tag{6.40}$$

$$|0\rangle = |0^0\rangle + |0^1\rangle + |0^2\rangle + \cdots \tag{6.41}$$

one then attempts to evaluate $(T_k^+|E\hat{1} + \hat{H}|T_l^+)$ to sufficiently high order to guarantee that the poles of primary interest are obtained accurately through a chosen order. If one is also interested in calculating residues that are accurate through some order, then the chosen operator manifold and reference state $|0\rangle$ must be taken to sufficient size and order to guarantee this. We return to the problem on how to choose $\{\mathbf{T}^+\}$ so as to determine primary poles and residues accurate through a chosen order in Section 6.C.4.

3. Hermiticity Questions

Earlier in this chapter, we noted that the question of the hermiticity of $(T_k^+|\hat{H}|T_l^+)$ had to be examined in individual cases (i.e., it was not automatically valid). When a perturbation expansion is used to determine the reference state, we may more explicitly state the conditions under which the matrix is hermitian by examining the difference between the (kl)th and the complex conjugate of the (lk)th element of the superoperator Hamiltonian. When this difference

$$(T_k^+|\hat{H}|T_l^+) - (T_l^+|\hat{H}|T_k^+)^* = \langle 0|[[T_l^+, T_k], H]|0\rangle \tag{6.42}$$

is equal to zero, the superoperator Hamiltonian is hermitian.

When the reference state $|0\rangle$ is determined through a certain order n in RSPT, ($|0\rangle_n = \sum_{i=0}^{n}|0^i\rangle$) the Schrödinger equation is solved through the same order:

$$H|0\rangle_n = E_n|0\rangle_n + 0(n+1) \tag{6.43}$$

where $E_n = \sum_{i=0}^{n} E^{(i)}$. Inserting (6.43) into (6.42) then states that Eq. (6.42) is zero through order n and that, as a result, the superoperator Hamiltonian matrix is hermitian through that same order. This theorem is quite useful for two reasons. First, it guarantees that the superoperator Hamiltonian matrix will have no accidental or spurious nonhermitian terms if it is properly calculated. Second, it is often easier to compute $(T_k^+|\hat{H}|T_l^+)$ than $(T_l^+|\hat{H}|T_k^+)$ (e.g., $(p^+q^+\alpha|\hat{H}|r^+)$ is easier than $(r^+|\hat{H}|p^+q^+\alpha)$ because the latter elements require that the Hamiltonian be commuted with $p^+q^+\alpha$). Thus, we can choose to calculate the "easier" matrix elements and to then obtain the others through hermiticity (i.e., by equating the complex conjugate of the former to the latter).

4. Operator Space Partitioning

We next go into more detail concerning the explicit evaluation of $\langle\langle A;B\rangle\rangle_E$ for $A = k^+$, $B^+ = l^+$ [referred to as the electron propagator (EP) or one-particle GF] and for $A = k^+l$, $B = i^+j$ [referred to as the polarization propagator (PP) or two-particle GF]. However, it remains for us to show one more approximation step that is often employed in searching for the poles of $(E\mathsf{S} + \hat{\mathsf{H}})^{-1}$ in Eq. (6.37). Because, according to Eq. (6.38), all elements of this inverse matrix possess poles at all of the E_j, it is possible to search for the desired poles by computing a single element or a submatrix of $(E\mathsf{S} + \hat{\mathsf{H}})^{-1}$. That is, if the operator manifold is partitioned into, say, two classes $\{\mathbf{T}_k^+\} = \{\mathbf{T}_a^+\} + \{\mathbf{T}_b^+\}$, then because $(E\mathsf{S} + \hat{\mathsf{H}})$ blocks into four submatrices

$$(E\mathsf{S} + \hat{\mathsf{H}}) = \begin{pmatrix} E\mathsf{S}_{aa} + \hat{\mathsf{H}}_{aa} & E\mathsf{S}_{ab} + \hat{\mathsf{H}}_{ab} \\ E\mathsf{S}_{ba} + \hat{\mathsf{H}}_{ba} & E\mathsf{S}_{bb} + \hat{\mathsf{H}}_{bb} \end{pmatrix} \tag{6.44}$$

one can solve for any element(s) of $(E\mathsf{S} + \hat{\mathsf{H}})^{-1}$ in terms of the above four submatrices. For example, it is easily shown that

$$(E\mathsf{S} + \hat{\mathsf{H}})_{aa}^{-1} = [(E\mathsf{S}_{aa} + \hat{\mathsf{H}}_{aa}) - (E\mathsf{S}_{ab} + \hat{\mathsf{H}}_{ab})(E\mathsf{S}_{bb} + \hat{\mathsf{H}}_{bb})^{-1}(E\mathsf{S}_{ba} + \hat{\mathsf{H}}_{ba})]^{-1} \tag{6.45}$$

Even if the space $\{\mathbf{T}_a^+\}$ includes a single element, if treated properly and to all orders, Eq. (6.45) will yield all the poles of $(E\mathsf{S} + \hat{\mathsf{H}})^{-1}$.

It is, of course, natural to wonder both why one would be interested in so partitioning $(E\mathsf{S} + \hat{\mathsf{H}})^{-1}$ and what this has to do with an approximation scheme for calculating $\langle\langle A;B\rangle\rangle_E$. It often turns out that if the sets $\{\mathbf{T}_a^+\}$ and $\{T_b^+\}$ are chosen properly, all the "off-diagonal" elements $(E\mathsf{S}_{ab} + \hat{\mathsf{H}}_{ab})$

(and hence $ES_{ba} + \hat{H}_{ba}$) contain only terms that are of first or higher order, whereas $(ES_{aa} + \hat{H}_{aa})$ contains zeroth- and (perhaps) higher-order terms. If, therefore, one restricts the search for poles to energy ranges in which $(ES_{bb} + \hat{H}_{bb})^{-1}$ is not close to being singular, then the term $(ES_{ab} + \hat{H}_{ab})(ES_{bb} + \hat{H}_{bb})^{-1}(ES_{ba} + \hat{H}_{ba})$ can be assumed to be of second or higher order. This restriction of the energy search range is often motivated by knowledge that the zeroth-order poles of $(ES_{aa} + \hat{H}_{aa})^{-1}$ are good approximations (e.g., through Koopmans' theorem for the IP) to the desired poles. If one is interested in calculating poles that are accurate to, say, second order, then the second- and higher-order pieces of $(ES_{ab} + \hat{H}_{ab})$ and the first- and higher-order pieces of $(ES_{bb} + \hat{H}_{bb})^{-1}$ can be neglected. In this way, one is often able to greatly simplify the calculation of certain poles of $\langle\langle A; B\rangle\rangle_E$ [those far from the singularities of $(ES_{bb} + \hat{H}_{bb})$ in the above example].

Given a choice of $\{\mathbf{T}_a^+\}$ and $\{\mathbf{T}_b^+\}$ that permits a pole (say E_j) of $\langle\langle A; B\rangle\rangle_E$ to be evaluated through a certain order, it still remains to examine whether the same partitioning will yield residues, which are given in Eq. (6.39), accurate to some chosen order. Thus, if $(\mathrm{B}^+|\mathbf{T}_a^+)$ and $(\mathbf{T}_a^+|A)$ are of zeroth and higher order, whereas $(B^+|\mathbf{T}_b^+)$ and $(\mathbf{T}_b^+|A)$ are of first and higher order, it is convenient to so partition $\{\mathbf{T}_k^+\}$ since the contributions to the desired residues can more easily be order analyzed. This point is made more clear when analyzing the residues of the polarization propagator in Section 6.E.4.

5. Nonperturbative Approaches

The perturbation theory approach to computing approximations to $\langle\langle A; B\rangle\rangle_E$ has been widely used with significant success. However, its fundamental premise (that U is "small") is known to break down under circumstances that are relatively widely appreciated (e.g., for $X^1\Sigma_g H_2$ at large internuclear distance, the contribution of the $1\sigma_u^2$ configuration can not be accurately represented by RSPT). For this reason, researchers have begun to explore the possibility of systematically calculating GFs in which the reference state $|0\rangle$ is taken to be of the MCSCF form. The MCSCF nature of $|0\rangle$ turns out to be very convenient in a GF analysis because the GBT results in hermiticity of certain blocks of the $(\mathbf{T}_k^+|\hat{H}|\mathbf{T}_l^+)$ matrix.

The primary formal difficulty that arises in implementing such MCSCF-based GFs has to do with developing systematic procedures for truncating (and perhaps partitioning) the $\{\mathbf{T}_k^+\}$ operator space. Because we have now lost the concept of order, we must turn to some other criterion for choosing an appropriate operator manifold. In the few developments of the MCSCF-based electron (Banerjee *et al.*, 1978) and polarization propagators (Yeager and Jørgensen, 1979; Dalgaard, 1980) that have been made to date, the

$\{\mathbf{T}_k^+\}$ manifold was chosen by examining the functions $\mathbf{T}_k^+|0\rangle$ and $\mathbf{T}_k|0\rangle$, which result from the application of the $\mathbf{T}_k^+$ operators to $|0\rangle$ as they occur in the superoperator binary product. Decisions were then made to guarantee that these functions contained all of the dominant singly and doubly excited configurations needed to yield proper orbital relaxation and electron correlation (or bond-breaking) effects, respectively. For example, the operator manifold $\{\mathbf{T}_k^+\} = \{r^+s, s^+r, r > s;\ |n\rangle\langle 0|, |0\rangle\langle n|\}$ has been used to express an MCSCF-based PP. The state projectors $|n\rangle\langle 0|$ and $|0\rangle\langle n|$ can be viewed, when they act on $|0\rangle$, as compact representations of the set of $\{\mathbf{T}_k^+\}$ operators given in Eq. (6.24). It is thus possible to choose another set of operators than the one of Eq. (6.24) to describe accurately the poles and residues of the PP. The decision to choose one truncation of $\{\mathbf{T}_k^+\}$ over the other is usually based upon considerations involving the dimension of the resulting $(\mathbf{T}^+|\hat{H}|\mathbf{T}^+)$ matrix and the ease of calculation of the requisite superoperator matrix elements. The first choice described above (involving the state projectors $|n\rangle\langle 0|, |0\rangle\langle n|$) seems to be especially promising because, as Dalgaard has demonstrated, this set of operators yields a PP whose poles and residues automatically guarantee equality between electric dipole transition moments computed within either the so-called length or velocity representations. This is especially convenient because one then has a continuous range [from the single-configuration time-dependent Hartree–Fock (TDHF) or random-phase approximation (RPA) through the present MCSCF case to the full CI] of PP approximations all of which preserve their length/velocity equivalence. Another reason for choosing the above set of operators for use as $\{\mathbf{T}_k^+\}$ lies in the fact that the resultant $(\mathbf{T}_k^+|\hat{H}|\mathbf{T}_l^+)$ matrix elements are no different than those arising in the original MCSCF calculation of $|0\rangle$ [e.g., $(r^+s|H||n\rangle\langle 0|)$ arises in $\langle 0|[\lambda,[H,S]]|0\rangle$ of Eq. (2.24)]. Also, if one were to consider the effect of an external one-electron perturbation on the MCSCF state $|0\rangle$, one would find the same operators $\{r^+s, s^+r, |n\rangle\langle 0|, |0\rangle\langle n|\}$ appearing naturally in the response of $|0\rangle$ to the external perturbation, as in coupled multiconfigurational HF.

6. Discussion

Because of the high research activity level on how to use an MCSCF reference in the GFs (EP and PP), it is not presently clear how to optimally choose truncated sets of $\{\mathbf{T}_k^+\}$ operators. It is likely that many workers will carry out test calculations involving many choices of the pertinent operator manifolds before this situation is improved. Moreover, questions concerning when and how to partition the resulting $(\mathbf{T}_k^+|\hat{H}|\mathbf{T}_l^+)$ matrix so as to reduce the dimension of the matrix whose poles are to be found remain unanswered for the case of an MCSCF reference function. Again, what is missing is some concept of order (or size or importance) in terms of which to make decisions about how to partition the operator manifold. It is our opinion that significant progress will be made on these important questions within the near future and that, as a result, MCSCF-based GF methods will become common tools in the quantum chemist's library.

Having given an introduction to the fundamental properties of GFs and to the techniques that are used to obtain GFs whose poles and residues are accurate to a chosen precision, we now move on to consider the commonly used EPs and PPs in some detail. We should mention that the resulting working equations arising in the EP and PP cases have also been derived through the so-called equations-of-motion (EOM) formalism (Schaefer and Miller, 1977, Chapter 9). This EOM formalism focuses on setting up the superoperator generalized eigenvalue problem of Eq. (6.33) and, as a result, is equivalent to the propagator development here. We do not enter into a closer discussion of the EOM development here because, for the EP and PP treated below, this tool offers no new insight or convenience.

D. THE ELECTRON PROPAGATOR

If we choose the A and B^+ operators to be of the one-electron addition form (r^+, s^+), then the GF $\langle\langle A;B\rangle\rangle_E$ is known as the EP:

$$\langle\langle r^+;s\rangle\rangle_E = (s^+|(E\hat{1} + \hat{H})^{-1}|r^+) \equiv G_{sr}(E) \tag{6.46}$$

This choice of A and B^+ is made because we are interested in studying primary ionization events [ionization potentials (Cederbaum, 1973; Pickup and Goscinski, 1973; Doll and Reinhardt, 1972; Purvis and Öhrn, 1974) and electron affinities (Simons and Smith, 1973; Jørgensen and Simons, 1975)], which may be reasonably described through acting with a single-electron operator (r^+ or r) on the reference state $|0\rangle$. To obtain computationally useful expressions for $G_{rs}(E)$ specific choices must be made for the reference state $|0\rangle$ and for the operator manifold $\{\mathbf{T}^+\}$ in Eq. (6.32). We describe a few of the most commonly employed choices of these quantities and the resulting GF.

1. Koopmans' Theorem

The simplest approximation to the EP is obtained by taking the reference state to be a single configuration HF wavefunction and the projection manifold to be

$$\{\mathbf{T}^+\} = \{\mathbf{T}_1^+\} = \{\boldsymbol{\alpha}, \mathbf{p}^+\} \tag{6.47}$$

The EP in Eq. (6.32) then reads

$$G_{sr} = ((s^+|\boldsymbol{\alpha}^+)(s^+|\mathbf{p}^+))\begin{pmatrix}(\boldsymbol{\alpha}^+|E\hat{I} + \hat{H}|\boldsymbol{\alpha}^+) & (\mathbf{p}^+|E\hat{I} + \hat{H}|\boldsymbol{\alpha}^+) \\ (\mathbf{p}^+|E\hat{I} + \hat{H}|\boldsymbol{\alpha}^+) & (\mathbf{p}^+|E\hat{I} + \hat{H}|\mathbf{p}^+\end{pmatrix}^{-1}\begin{pmatrix}(\boldsymbol{\alpha}^+|r^+) \\ (\mathbf{p}^+|r^+)\end{pmatrix} \tag{6.48}$$

The matrix elements appearing in Eq. (6.48) can easily be evaluated because of the single-determinant nature of $|0\rangle$:

$$(s^+|m^+) = \langle 0|[s, m^+]_+|0\rangle = \delta_{sm} \tag{6.49}$$

$$(s^+|\beta^+) = \delta_{s\beta} \tag{6.50}$$

$$\begin{aligned}(\beta^+|E\hat{I} + \hat{H}|\gamma^+) &= \delta_{\beta\gamma}E + \langle 0|[\beta, [H, \gamma^+]]_+|0\rangle \\ &= E\delta_{\beta\gamma} + h_{\beta\gamma} + \sum_{kt}\langle \beta k||\gamma t\rangle\langle 0|k^+t|0\rangle \\ &= (E + \varepsilon_\gamma)\delta_{\beta\gamma}\end{aligned} \tag{6.51}$$

$$(m^+|E\hat{I} + \hat{H}|n^+) = (E + \varepsilon_m)\delta_{mn} \tag{6.52}$$

$$(m^+|E\hat{I} + \hat{H}|\alpha^+) = 0 \tag{6.53}$$

Here $\{\varepsilon_i\}$ denotes HF orbital energies. Using these results, Eq. (6.48) may be expressed as

$$G_{sr} = \sum_m \frac{\delta_{sm}\delta_{rm}}{E + \varepsilon_s} + \sum_\gamma \frac{\delta_{s\gamma}\delta_{r\gamma}}{E + \varepsilon_s} \tag{6.54}$$

By comparing the spectral representation of the GF in Eq. (6.6) with Eqs. (6.48) and (6.54) we see that the pole of Eq. (6.54) at $E = -\varepsilon_m$ represents an approximation to the electron affinity, while the pole at $E = -\varepsilon_\gamma$ corresponds to an ionization potential. The residue (the square of the transition amplitude) at $E = -\varepsilon_m$ is $\delta_{sm}\delta_{rm}$, while the residue at $E = -\varepsilon_\gamma$ is $\delta_{s\gamma}\delta_{r\gamma}$. All transition amplitudes corresponding to primary ionization events thus become equal to unity at this level of approximation. The above result expresses the EP analog of Koopmans' theorem. To go beyond Koopmans' theorem, better choices must be made for the reference state and operator manifold.

2. Rayleigh–Schrödinger Order Analysis

As discussed in Section A, RSPT has been widely used to develop systematic approximations to $\mathrm{G}(E)$. Here the unperturbed Hamiltonian H^0 is taken to be the HF Hamiltonian [Eq. (3.34)] and the orthonormal basis spin-orbitals are HF spin-orbitals having orbital energies ε_i:

$$H^0 = \sum_j \varepsilon_j j^+ j \tag{6.55}$$

The reference function $|0\rangle$ can be expressed, as in Chapter 3, as a perturbation series in powers of the residual electron–electron interaction. The EP is then used to describe the primary ionization events consistent through a certain order by expanding the reference state in powers of the fluctuation potential and by choosing the projection manifold of Eq. (6.23) to be sufficently large, the meaning of which will be discussed later. In this section, we show how to determine the primary ionization events consistent through zeroth, first, second, and third order. To do so, it proves sufficient to consider the truncated manifold

$$\{\mathbf{T}^+\} = \{\mathbf{T}_1^+;\mathbf{T}_3^+\} \tag{6.56}$$

This conclusion is by no means obvious but should become clear shortly. One must, in principle, examine the interaction between $\mathbf{T}_1^+$, $\mathbf{T}_3^+$ and the $\mathbf{T}_5^+$, $\mathbf{T}_7^+$, etc., operators to conclude that these higher operators have no effect on the poles describing the primary ionization event through third order (Redmon *et al.*, 1975).

With the above choice of the projection manifold, the EP of Eq. (6.32) takes the form

$$G_{sr}(E) = ((s^+|\mathbf{T}_1^+)(s^+|\mathbf{T}_3^+))\begin{pmatrix} \mathsf{A} & \mathsf{C} \\ \mathsf{C}^\mathrm{T} & \mathsf{M}\end{pmatrix}^{-1}\begin{pmatrix}(\mathbf{T}_1^+|r^+) \\ (\mathbf{T}_3^+|r^+)\end{pmatrix} \tag{6.57}$$

where the matrices in Eq. (6.57) are defined as

$$\mathsf{A} = (\mathbf{T}_1^+|E\hat{1} + \hat{H}|\mathbf{T}_1^+) \tag{6.58}$$

$$\mathsf{C} = (\mathbf{T}_3^+|E\hat{1} + \hat{H}|\mathbf{T}_1^+) \tag{6.59}$$

$$\mathsf{M} = (\mathbf{T}_3^+|E\hat{1} + \hat{H}|\mathbf{T}_3^+) \tag{6.60}$$

The poles of the GF are determined entirely by the inverse matrix of Eq. (6.57). Since our interest is in describing the primary ionization events, we partition the inverse matrix as in Eqs. (6.44) and (6.45) with $\mathbf{T}_a^+ = \mathbf{T}_1^+$, and $\mathbf{T}_b^+ = \mathbf{T}_3^+$. We then determine the poles that describe the primary ionization events from the partitioned form of the inverse matrix

$$\mathsf{P}^{-1}(E) = (\mathsf{A} - \mathsf{C}^\mathrm{T}\mathsf{M}^{-1}\mathsf{C})^{-1} \tag{6.61}$$

By using $H = H^0 + U$ [see Eqs. (3.34) and (3.35)] and $|0\rangle = |0^0\rangle + |0^1\rangle + \ldots$, we may carry out a detailed order analysis of each of the four matrices A, B, C, and M. For example, we write A as

$$\mathsf{A} = \sum_{i=0}^{\infty} \mathsf{A}_i \tag{6.62}$$

where the label i indicates the order of the contributions to the A matrix.

Below we show all the contributions to the matrix A through third order:

$$(\mathsf{A}_0)_{jk} = E\langle 0^0|[j,k^+]_+|0^0\rangle + \langle 0^0|[j,[H^0,k^+]]_+|0^0\rangle = (E+\varepsilon_j)\delta_{kj} \qquad (6.63)$$

$$\mathsf{A}_1 = \mathsf{A}_2 = 0 \qquad (6.64)$$

$$(\mathsf{A}_3)_{jk} = \langle 0^1|[j,[U,k^+]]_+|0^1\rangle + \langle 0^2|[j,[U,k^+]_+|0^0\rangle + \langle 0^0|[j,[U,k^+]]_+|0^2\rangle = \sum_{il}\langle ji||kl\rangle\left(\sum_{\substack{\alpha>\beta\\ p}} K^{pi}_{\alpha\beta}K^{pl}_{\alpha\beta} - \sum_{\substack{q>p\\ \alpha}} K^{pq}_{\alpha l}K^{pq}_{\alpha i}\right) + \sum_{p\delta}(\langle j\delta||kp\rangle + \langle jp||k\delta\rangle)K^p_\delta \qquad (6.65)$$

where the perturbation theory correlation coefficients are given in Eqs. (3.53) and (3.55). It should be noticed that both A_1 and A_2 are identically zero. This fact will be shown to lead to the conclusion that the Koopmans' theorem approximation to $\mathsf{G}(E)$ is accurate through first order. We also list below all of the matrix elements of the C and M matrices, which are required to evaluate $\mathsf{P}(E)$, and hence to obtain poles of $\mathsf{G}(E)$, through third order:

$$\mathsf{C}_0 = 0 \qquad (6.66)$$

$$(\mathsf{C}_1)_{pq\alpha,j} = \langle 0^0|[\alpha^+qp,[U,j^+]]_+|0^0\rangle = \langle pq||j\alpha\rangle \qquad (6.67)$$

$$(\mathsf{C}_1)_{\alpha\beta m,j} = -\langle\alpha\beta||jm\rangle \qquad (6.68)$$

$$(\mathsf{C}_2)_{pq\alpha,j} = \tfrac{1}{2}\sum_{\gamma\delta}\langle i\alpha||\delta\gamma\rangle K^{qp}_{\delta\gamma} + \sum_{\gamma m}[\langle im||\gamma p\rangle K^{qm}_{\alpha\gamma} - \langle im||\gamma q\rangle K^{pm}_{\alpha\gamma}] \qquad (6.69)$$

$$(\mathsf{C}_2)_{\alpha\beta m,j} = -\tfrac{1}{2}\sum_{pq}\langle im||pq\rangle K^{pq}_{\alpha\beta} + \sum_{\gamma p}[\langle i\gamma||p\alpha\rangle K^{mp}_{\beta\gamma} - \langle i\gamma||p\beta\rangle K^{mp}_{\alpha\gamma}] \qquad (6.70)$$

$$(\mathsf{M}_0)_{nm\alpha,qp\beta} = \delta_{nq}\delta_{mp}\delta_{\alpha\beta}(E+\varepsilon_n+\varepsilon_m-\varepsilon_\alpha) \qquad (6.71)$$

$$(\mathsf{M}_0)_{\delta\gamma p,\alpha\beta q} = \delta_{\delta\alpha}\delta_{\gamma\beta}\delta_{pq}(E+\varepsilon_\delta+\varepsilon_\gamma-\varepsilon_p) \qquad (6.72)$$

$$(\mathsf{M}_0)_{nm\alpha,\gamma\delta p} = 0 \qquad (6.73)$$

$$(\mathsf{M}_1)_{nm\alpha,qp\beta} = -\delta_{qn}\langle m\beta||p\alpha\rangle - \delta_{pm}\langle n\beta||q\alpha\rangle + \delta_{qm}\langle n\beta||p\alpha\rangle + \delta_{\alpha\beta}\langle mn||pq\rangle + \delta_{pn}\langle m\beta||q\alpha\rangle \qquad (6.74)$$

$$(\mathsf{M}_1)_{\delta\gamma p,\alpha\beta q} = \delta_{\gamma\beta}\langle\delta q||\alpha p\rangle + \delta_{\delta\alpha}\langle\gamma q||\beta p\rangle - \delta_{\gamma\alpha}\langle\delta q||\beta p\rangle + \delta_{pq}\langle\delta\gamma||\beta\alpha\rangle - \delta_{\delta\beta}\langle\gamma q||\alpha p\rangle \qquad (6.75)$$

$$(\mathsf{M}_1)_{nm\alpha,\delta\gamma p} = 0 \qquad (6.76)$$

a. *Pole Structure through Second Order*

The poles of the EP consistent through zeroth order are determined by including all zeroth-order terms in $\mathsf{P}(E)$ [Eq. (6.61)]. Since C contains no zeroth-order contributions, we find that

$$\mathsf{P}_0(E) = \mathsf{A}_0 \qquad (6.77)$$

which is the Koopmans' theorem result once again. A determination of $\mathsf{P}(E)$ through first order would not contain any more terms than are already in A_0 since $\mathsf{A}_1 = 0$ and $\mathsf{C}_0 = 0$. The lowest-order correction to A_0 given by $\mathsf{C}^{\mathrm{T}}\mathsf{M}^{-1}\mathsf{C}$ occurs in second order. In reaching this conclusion, we used the fact that the order of a term that is a product of several matrices is determined by adding up the individual orders of the matrices appearing in the term. The term $\mathsf{C}_0^{\mathrm{T}}\mathsf{M}_0^{-1}\mathsf{C}_1$ would thus have been of first order if C_0 had not vanished. An EP that contains only the A_0 matrix is identical to the EP obtained in Section 6.D.1 and results in Koopmans' theorem–level estimates of electron affinities and ionization potentials. The success of using Koopmans' theorem to assign peaks in photoelectron spectra relies on the fact that corrections to Koopmans' theorem first appear in second order.

Proceeding now to compute all terms in Eq. (6.61) through second order, we find

$$\mathsf{P}_2(E) = \mathsf{A}_0 - \mathsf{C}_1^{\mathrm{T}}\mathsf{M}_0^{-1}\mathsf{C}_1 \tag{6.78}$$

since $\mathsf{A}_2 = 0$. In all of the matrices in $\mathsf{P}_2(E)$, only the zeroth-order part of the reference state $|0^0\rangle$ contributes, as can be seen by examining Eqs. (6.66)–(6.76). Inserting the expressions for the individual matrix elements of C_1 and M_0 given in Eqs. (6.67), (6.68), (6.71)–(6.73) into Eq. (6.78) gives explicit expressions for the elements of $\mathsf{P}_2(E)$:

$$[\mathsf{P}_2(E)]_{jk} = (E + \varepsilon_j)\delta_{jk} - \sum_{\substack{p>q\\ \alpha}} \frac{\langle j\alpha||pq\rangle\langle pq||k\alpha\rangle}{E - \varepsilon_\alpha + \varepsilon_p + \varepsilon_q} - \sum_{\substack{p\\ \alpha>\beta}} \frac{\langle jp||\alpha\beta\rangle\langle\alpha\beta||kp\rangle}{E + \varepsilon_\alpha + \varepsilon_\beta - \varepsilon_p} \tag{6.79}$$

Such second-order EPs have been used (Doll and Reinhardt, 1972; Purvis and Öhrn, 1974) to compute atomic and molecular ionization potentials, electron affinities, and even electron–atom shape resonance positions and lifetimes with some success. Based upon the experience gained to date, however, we cannot expect the accuracy of this approach to be better than ± 0.5 eV, even for systems that are described reasonably well by a single-configuration reference function. Often, this numerical accuracy is not satisfactory and hence the above formalism must be advanced to higher order (or replaced by another development that does not depend upon the Rayleigh–Schrödinger order concept). An example of such a second-order EP calculation is given in Problem 6.1.

b. Physical Interpretation

The physical interpretation of terms arising in $\mathsf{C}_1^{\mathrm{T}}\mathsf{M}_0^{-1}\mathsf{C}_1$ in terms of orbital relaxation and electron pair correlation effects has been carried out by several workers. To give some feeling for the physical content of the terms in $(\mathsf{P}_2)_{ij}$, we examine the diagonal $i = j = t$ term, which would be expected to be the dominant contributor in the case of adding an electron to spin-orbital ϕ_t. Then, through second order, from Eq. (6.79) we have

$$(P_2)_{tt} = \varepsilon_t + E + \sum_{\alpha>\beta,m} \frac{|\langle tm||\alpha\beta\rangle|^2}{\varepsilon_m + \varepsilon_t - \varepsilon_\alpha - \varepsilon_\beta}$$
$$- \sum_{q\neq t;\alpha} \frac{|\langle t\alpha||tq\rangle|^2}{\varepsilon_q - \varepsilon_\alpha} - \sum_{\substack{p>q\neq t\\ \alpha}} \frac{|\langle t\alpha||pq\rangle|^2}{\varepsilon_p + \varepsilon_q - \varepsilon_t - \varepsilon_\alpha} \tag{6.80}$$

Clearly this term will vanish (G will have a pole) near $E = -\varepsilon_t$, which is the Koopmans' theorem estimate. The correction to Koopmans' theorem expressed in the three sums occurring above can be given physical meaning in the following manner. The second sum gives the orbital relaxation contribution to the ion–neutral energy difference. By expanding the HF orbitals of the ion in terms of those of the neutral and then computing the ion's energy with these orbitals correct through second order one could derive this term within a wavefunction picture (Pickup and Goscinski, 1973). The fact that this sum has an orbital energy denominator involving only a single orbital excitation energy ($\varepsilon_q - \varepsilon_\alpha$) has to do with the fact that, in a configuration interaction language, this term arises from single spin-orbital excitations ($\phi_\alpha \rightarrow \phi_q$). The numerator $|\langle t\alpha||tq\rangle|^2$ can be identified as the square of the perturbation matrix element coupling orbitals ϕ_α and ϕ_q. The perturbation is the coulomb and exchange potential caused by the electron that has been added into ϕ_t. The third sum in Eq. (6.80) gives the approximate correlation energy of an electron in ϕ_t with the remaining N electrons (in ϕ_α) and hence has to do with double excitations ($\phi_t\phi_\alpha \rightarrow \phi_p\phi_q$), which would arise in a CI description of such pair correlations. Finally, the first sum describes the changes in the correlation energies between pairs of orbitals ϕ_α, ϕ_β due to the fact that spin-orbital ϕ_t is occupied in the ion (and hence unavailable for correlating ϕ_α and ϕ_β), but was not occupied in the neutral parent molecule.

c. *Third-Order Analysis of Pole Structure*

To obtain the expression for $\mathsf{P}(E)$ that contains all terms through third order (Simons and Smith, 1973; Jørgensen and Simons, 1975; Cederbaum, 1973), we introduce the matrices given explicitly in Eqs. (6.63)–(6.76) into Eq. (6.61) and neglect the fourth-order terms. This allows us to write Eq. (6.61) as

$$\mathsf{P}(E) = \mathsf{A}_0 + \mathsf{A}_3 - \mathsf{C}_1^{\mathrm{T}}(\mathsf{M}_0 + \mathsf{M}_1)^{-1}\mathsf{C}_1 - \mathsf{C}_1^{\mathrm{T}}(\mathsf{M}_0 + \mathsf{M}_1)^{-1}\mathsf{C}_2$$
$$- \mathsf{C}_2^{\mathrm{T}}(\mathsf{M}_0 + \mathsf{M}_1)^{-1}\mathsf{C}_1 \tag{6.81}$$

The inverse matrix $(\mathsf{M}_0 + \mathsf{M}_1)^{-1}$ can further be decomposed into orders by performing the expansion

$$(\mathsf{M}_0 + \mathsf{M}_1)^{-1} = \mathsf{M}_0^{-1} - \mathsf{M}_0^{-1}\mathsf{M}_1\mathsf{M}_0^{-1} + \mathsf{M}_0^{-1}\mathsf{M}_1\mathsf{M}_0^{-1}\mathsf{M}_1\mathsf{M}_0^{-1} + \cdots \tag{6.82}$$

which can be used in Eq. (6.81) to identify the terms to keep through third order:

$$\mathsf{P}_3(E) = \mathsf{A}_0 + \mathsf{A}_3 - \mathsf{C}_1^{\mathsf{T}}\mathsf{M}_0^{-1}\mathsf{C}_1 + \mathsf{C}_1^{\mathsf{T}}\mathsf{M}_0^{-1}\mathsf{M}_1\mathsf{M}_0^{-1}\mathsf{C}_1 - \mathsf{C}_1^{\mathsf{T}}\mathsf{M}_0^{-1}\mathsf{C}_2 - \mathsf{C}_2^{\mathsf{T}}\mathsf{M}_0^{-1}\mathsf{C}_1 \tag{6.83}$$

At this third-order level of approximation, the EP has been successfully applied to a large number of inorganic and organic molecules. The ionization potentials (Von Niessen, *et al.*, 1979) and electron affinities (Simons, 1977) thus obtained are usually reliable to within ± 0.3 eV.

d. Diagrammatic Analysis

The derivation of the EP consistent through a certain order may alternatively be performed in a way very similar to that used in MBPT to express the state energy and wavefunction. As in MBPT, the result is expressed in terms of a set of diagrams. In this section we give the results of performing such a diagrammic perturbation analysis of the EP. The contribution to $\mathsf{P}(E)$ beyond the $(E + \varepsilon_i)\delta_{ij}$ is, in the diagrammatic analysis, referred to as the self-energy or optical potential matrix $\Sigma(E)$. The self-energy matrix $\Sigma(E)$ in a given order n is expressed in terms of a set of Hugenholtz diagrams. The diagrams, which enter in order n, are determined by applying the rules in Table I of Chapter 3, with rule 3 modified such that when one is connecting lines each diagram has to have one incoming and one outgoing line. The translation of a Hugenholtz diagram into an algebraic expression is, as in MBPT, performed by translating the Hugenholtz diagram into one of its equivalent Brandow diagrams (Fig. 6.1; see Section 3.G). The algebraic expression for the Brandow GF diagram is obtained by applying the rules of Table II of Chapter 3 with rule 3 modified such that an energy parameter equal to $(-1)^h E$ is added to each factor in the denominator $\sum_\alpha \varepsilon_\alpha - \sum_p \varepsilon_p$ if

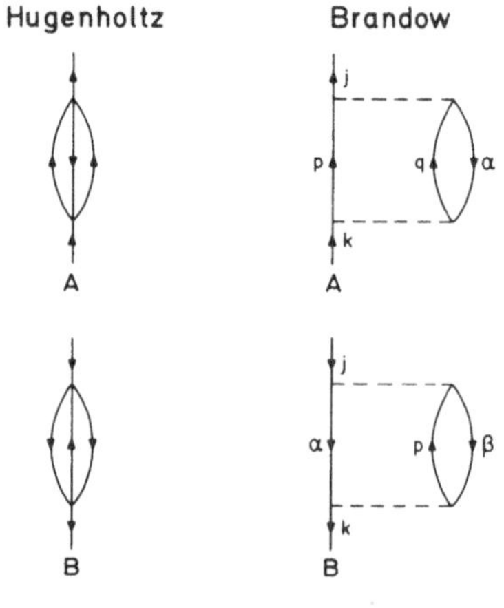

FIG. 6.1. All Hugenholtz and Brandow second-order self-energy diagrams.

the sum of the number of internal hole and particle lines is an odd integer. Here h is the number of internal hole lines in the diagram. Lines are counted as internal only if they lie between the vertices from which the GF's two free lines originate. If the GF's two free lines start at the same point, only those hole lines that exist horizontal to this point are counted. For example, diagrams A and C of Fig. 6.2 contain one and zero internal line, respectively. Diagram G of Fig. 6.2 has one internal hole line and thus each denominator would get a $-E$ factor added in.

If H^0 is taken to be the HF Hamiltonian, then, as in MBPT, all diagrams containing the loop structure cancel with the corresponding diagrams having the potential symbol in the same location. No first-order diagrams then enter in the diagrammatic perturbation analysis. In second order only the two diagrams displayed in Fig. 6.1 enter. To obtain some experience in applying the rules in Table II of Chapter 3, we list the analytical expressions for these two diagrams:

$$A = \sum_{\substack{pq \\ \alpha}} (\tfrac{1}{2})^1 (-1)^{1+1} \frac{\langle j\alpha || pq\rangle \langle pq || k\alpha\rangle}{-E + \varepsilon_\alpha - \varepsilon_p - \varepsilon_q} \tag{6.84}$$

$$B = \sum_{\substack{\alpha\beta \\ p}} (\tfrac{1}{2})^1 (-1)^{2+1} \frac{\langle \alpha\beta || jp\rangle \langle kp || \alpha\beta\rangle}{E + \varepsilon_\alpha + \varepsilon_\beta - \varepsilon_p} \tag{6.85}$$

The second-order contribution to the EP given above is, of course, identical to the one derived in Eq. (6.79).

In Fig. 6.2, we list the nonvanishing third-order self-energy diagrams. These may, of course, also be identified with corresponding terms of the

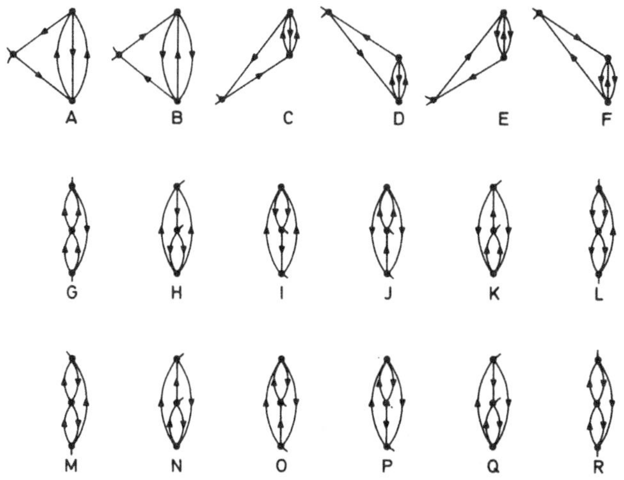

FIG. 6.2. All third-order Hugenholtz self-energy diagrams.

third-order expression of $\mathsf{P}_3(E)$ in Eq. (6.83). The sum of diagrams A and B represents the first two terms of A_3 in Eq. (6.65), while diagrams C–F may be identified as orginating from the last term of Eq. (6.65). The term $\mathsf{C}_1^{\mathsf{T}}\mathsf{M}_0^{-1}\mathsf{M}_1\mathsf{M}_0^{-1}\mathsf{C}_1$ gives rise to G, L, M, and R, while H–K and N–Q may be shown to originate from $\mathsf{C}_2^{\mathsf{T}}\mathsf{M}_0^{-1}\mathsf{C}_1$ and $\mathsf{C}_1^{\mathsf{T}}\mathsf{M}_0^{-1}\mathsf{C}_2$. The analytical third-order expression thus gives a compact representation of the diagrams in Fig. 6.2.

E. THE POLARIZATION PROPAGATOR

1. Introduction

If we choose the operators A and B^+ both to be the electric dipole moment operator $\mathbf{r}$, then the spectral representation of the resulting GF reads

$$\langle\langle \mathbf{r};\mathbf{r}\rangle\rangle_E = \lim_{\eta\to 0}\sum_m \left\{-\frac{\langle 0|\mathbf{r}|m\rangle\langle m|\mathbf{r}|0\rangle}{E - E_m + E_0 + i\eta} + \frac{\langle 0|\mathbf{r}|m\rangle\langle m|\mathbf{r}|0\rangle}{E + E_m - E_0 - i\eta}\right\} \tag{6.86}$$

The residue at the pole $E = \pm(E_m - E_0)$ contains the transition dipole matrix element between the states $|0\rangle$ and $|m\rangle$,

$$\langle 0|\mathbf{r}|m\rangle = \sum_{js}(\mathbf{r})_{js}\langle 0|j^+s|m\rangle \tag{6.87}$$

where

$$(\mathbf{r})_{js} \equiv \langle\phi_j|\mathbf{r}|\phi_s\rangle \tag{6.88}$$

Since $\mathbf{r}$ is a number-conserving operator, the reference state $|0\rangle$ and the state $|m\rangle$ must contain the same number N of electrons. The poles of this so-called polarization propagator (PP) thus occur at the excitation energies $E = \pm(E_m - E_0)$ of the system described by $|0\rangle$, while the corresponding residues give the squares of the electric dipole transition moments $|\langle 0|\mathbf{r}|m\rangle|^2$.

The real part of the above GF may be expressed by combining terms over a common denominator as

$$Re\langle\langle \mathbf{r};\mathbf{r}\rangle\rangle_E = -\sum_m \frac{2(E_m - E_0)|\langle 0|\mathbf{r}|m\rangle|^2}{E^2 - (E_m - E_0)^2} \tag{6.89}$$

which is identical to the conventional expression for the frequency-dependent polarizability tensor (the frequency being represented by E).

To get some experience in using the PP to express second-order frequency-dependent and -independent properties and to indicate some problems that may appear when using the PP in finite basis set calculations, we now derive alternative but formally equivalent expressions for the frequency-dependent polarizability. We may employ Eq. (6.10) to reexpress the propagator $\langle\langle \mathbf{r};\mathbf{r}\rangle\rangle_E$ as

$$E\langle\langle \mathbf{r};\mathbf{r}\rangle\rangle_E = \langle 0|[\mathbf{r},\mathbf{r}]|0\rangle + \langle\langle [\mathbf{r},H];\mathbf{r}\rangle\rangle_E = i\langle\langle \mathbf{p};\mathbf{r}\rangle\rangle_E \tag{6.90}$$

where we have used (in a.u.)

$$[\mathbf{r},H] = i\mathbf{p} \tag{6.91}$$

For $E \neq 0$, $iE^{-1}\langle\langle \mathbf{p};\mathbf{r}\rangle\rangle_E$ may alternatively be used to calculate the frequency-dependent polarizability. Near $E = 0$, however, we expect $iE^{-1}\langle\langle \mathbf{p};\mathbf{r}\rangle\rangle_E$ (which should, in principle, equal $\langle\langle \mathbf{r};\mathbf{r}\rangle\rangle_E$) to have difficulty in finite-basis-set calculations because of the explicit appearance of the E^{-1} factor. That is, unless $\langle\langle \mathbf{p};\mathbf{r}\rangle\rangle_E$, as a calculated function of E, is proportional to E near $E = 0$, one might obtain incorrect behavior of $iE^{-1}\langle\langle \mathbf{p};\mathbf{r}\rangle\rangle_E$ here.

Applying Eq. (6.10) once more to Eq. (6.90) gives

$$E\langle\langle \mathbf{p};\mathbf{r}\rangle\rangle_E = \langle 0|[\mathbf{p},\mathbf{r}]|0\rangle + \langle\langle \mathbf{p};[H,\mathbf{r}]\rangle\rangle_E = \langle 0|[\mathbf{p},\mathbf{r}]|0\rangle - i\langle\langle \mathbf{p};\mathbf{p}\rangle\rangle_E \tag{6.92}$$

Using the second-quantized forms for $\mathbf{p}$ and $\mathbf{r}$, we can explicitly calculate the commutator in Eq. (6.92) as

$$\begin{aligned}[\mathbf{p},\mathbf{r}] &= \sum_{ijkl} (\mathbf{p})_{ij}(\mathbf{r})_{kl}(\delta_{jk}i^+l - \delta_{il}k^+j) \\ &= \sum_{jl} [(\mathbf{pr})_{jl} - (\mathbf{rp})_{jl}]j^+l \\ &= -i\sum_{jl} \mathsf{I}_{jl}j^+l \end{aligned} \tag{6.93}$$

where I is the unit tensor operator whose elements are $\mathsf{I}_{jl} = \delta_{jl}\mathsf{I}$, and

$$(\mathsf{I})^{xx} = (\mathsf{I})^{yy} = (\mathsf{I})^{zz} = 1, \qquad (\mathsf{I})^{xy} = (\mathsf{I})^{xz} = (\mathsf{I})^{yz} = 0 \tag{6.94}$$

Clearly Eq. (6.93) is valid only if the basis set is complete so that we can write $(\mathbf{pr})_{jl} - (\mathbf{rp})_{jl} = (\mathbf{pr} - \mathbf{rp})_{jl} = -i\delta_{jl}\mathsf{I}$. We may now rewrite Eq. (6.92) to obtain one further expression for the frequency-dependent polarizability

$$\langle\langle \mathbf{r};\mathbf{r}\rangle\rangle_E = \frac{1}{E^2}(N\mathsf{I} + \langle\langle \mathbf{p};\mathbf{p}\rangle\rangle_E) \tag{6.95}$$

where the number N of electrons in $|0\rangle$ arises by evaluating $\mathsf{I}\sum_j \langle 0|j^+j|0\rangle$. As with $i\langle\langle \mathbf{p}_j\mathbf{r}\rangle\rangle_E E^{-1}$, finite-basis-set calculations of this form for the polarizability through the propagator $\langle\langle \mathbf{p};\mathbf{p}\rangle\rangle_E$ would be expected to have difficulty near $E = 0$ because the small-E portion of $\langle\langle \mathbf{p},\mathbf{p}\rangle\rangle_E$, which should exactly cancel the $N\mathsf{I}$ factor may, in a finite basis, not lead to exact cancellation.

We have now discussed how frequency-dependent polarizabilities can be obtained directly from the PP once a closed algebraic equation for $\langle\langle \mathbf{r};\mathbf{r}\rangle\rangle_E$ is found. Other second-order properties can equally well be determined by replacing A and B^+ with other one-electron operators (e.g., the dominant term in the indirect nuclear spin–spin coupling constant results when A and B^+ are taken to be the Fermi contact Hamiltonian).

In the following, we concentrate on how approximate closed expressions may be obtained for the $\langle\langle \mathbf{r};\mathbf{r}\rangle\rangle_E$ form of the polarization propagator. From our treatment of the PP it should become clear how to determine other second-order properties corresponding to other choices for A and B^+.

2. The Single-Configuration TDHF Approximation

In a simple and very commonly used approximation to the PP, the reference state $|0\rangle$ is chosen to be a single-configuration (but not necessarily single determinant) HF wavefunction. The operator manifold $\{\mathbf{T}^+\}$ then is taken as the set of particle–hole excitation and deexcitation operators used for optimizing the reference state:

$$\mathbf{T}^+ = \{\mathbf{T}_2^+\} = \{\mathbf{Q}^+, \mathbf{Q}\} = \{m^+\alpha, \alpha^+ m; m \quad \alpha\} \tag{6.96}$$

With these choices, the propagator takes the form (as expressed in Eq. (6.32)]

$$\langle\langle \mathbf{r};\mathbf{r}\rangle\rangle_E = ((\mathbf{r}|\mathbf{Q}^+)(\mathbf{r}|\mathbf{Q}))\begin{pmatrix} (\mathbf{Q}^+|E\hat{1}+\hat{H}|\mathbf{Q}^+) & (\mathbf{Q}^+|E\hat{1}+\hat{H}|\mathbf{Q}) \\ (\mathbf{Q}|E\hat{1}+\hat{H}|\mathbf{Q}^+) & (\mathbf{Q}|E\hat{1}+\hat{H}|\mathbf{Q}) \end{pmatrix}^{-1} \begin{pmatrix} (\mathbf{Q}^+|\mathbf{r}) \\ (\mathbf{Q}|\mathbf{r}) \end{pmatrix} \tag{6.97}$$

Since the one-particle density matrix is diagonal for the chosen HF reference state, we have

$$(\mathbf{Q}|\mathbf{Q}^+) = (\mathbf{Q}^+|\mathbf{Q}) = \{\langle 0[s^+\beta, r^+\alpha]|0\rangle\} = \{0\} \tag{6.98}$$

and

$$S_{r\alpha,s\beta} \equiv (\mathbf{Q}^+|\mathbf{Q}^+)_{r\alpha,s\beta} = \langle 0|[\alpha^+ r, s^+\beta]|0\rangle = \delta_{rs}\delta_{\alpha\beta}(\nu_\alpha - \nu_r) \tag{6.99}$$

and similarly

$$(\mathbf{r}|\mathbf{Q}^+)_{s\beta} = (\nu_\beta - \nu_s)(\mathbf{r})_{\beta s} \tag{6.100}$$

where ν_k is the occupancy of spin-orbital ϕ_k. Equation (6.97) may be written in more compact notation as

$$\langle\langle \mathbf{r};\mathbf{r}\rangle\rangle_E = ((\mathbf{r}|\mathbf{Q}^+)(\mathbf{r}|\mathbf{Q}))\begin{pmatrix} \mathsf{S}E + \mathsf{A}_{11} & \mathsf{B}_{11} \\ \mathsf{B}_{11} & -\mathsf{S}E + \mathsf{A}_{11} \end{pmatrix}^{-1} \begin{pmatrix} (\mathbf{Q}^+|\mathbf{r}) \\ (\mathbf{Q}|\mathbf{r}) \end{pmatrix} \tag{6.101}$$

where the matrices A_{11} and B_{11} are identical to those defined in connection with the MCSCF orbital optimization in Eqs. (2.29) and (2.30) except that $|0\rangle$ is taken here to be the single-configuration HF function. These matrix elements are evaluated explicitly in Problem 5.2 and used in Problem 6.2 to carry out a PP calculation.

Equation (6.101) is said to express the time-dependent Hartree–Fock (TDHF) or the random phase approximation (RPA) to the PP (Jørgensen, 1975). The TDHF (or RPA) approximation has been derived in a variety of ways, each of which tends to stress a certain aspect or point of view. In the following, we examine the physical content of the TDHF approximation and try to point out various consequences of using it for calculating the frequency-dependent polarizability, oscillator strengths, and excitation energies.

a Pole and Residue Analysis

We now demonstrate how the TDHF propagator may be transformed to a spectral form similar to the one appearing in Eq. (6.6). The poles of Eq. (6.101) can be determined through solving the nonhermitian eigenvalue problem

$$\begin{pmatrix} \mathsf{A}_{11} & \mathsf{B}_{11} \\ \mathsf{B}_{11} & \mathsf{A}_{11} \end{pmatrix} \begin{pmatrix} \mathbf{Z} \\ \mathbf{Y} \end{pmatrix} = E \begin{pmatrix} \mathsf{S} & 0 \\ 0 & -\mathsf{S} \end{pmatrix} \begin{pmatrix} \mathbf{Z} \\ \mathbf{Y} \end{pmatrix} \tag{6.102}$$

whose dimension is the sum of both the number of nonredundant particle–hole and hole–particle operators. The solution of Eq. (6.102) may alternatively be obtained through performing a series of transformations involving matrices of only the dimension of the particle–hole operators (Linderberg and Öhrn, 1977; Jørgensen, Olsen, and Yeager, 1981). To achieve this reduction in the matrix dimension, we first write Eq. (6.102) in component form as

$$\mathsf{A}_{11}\mathbf{Z} + \mathsf{B}_{11}\mathbf{Y} = E\mathsf{S}\mathbf{Z} \tag{6.103}$$

$$\mathsf{B}_{11}\mathbf{Z} + \mathsf{A}_{11}\mathbf{Y} = -E\mathsf{S}\mathbf{Y} \tag{6.104}$$

Successively adding and subtracting the above two equations gives

$$(\mathsf{A}_{11} + \mathsf{B}_{11})(\mathbf{Z} + \mathbf{Y}) = E\mathsf{S}(\mathbf{Z} - \mathbf{Y}) \tag{6.105}$$

$$(\mathsf{A}_{11} - \mathsf{B}_{11})(\mathbf{Z} - \mathbf{Y}) = E\mathsf{S}(\mathbf{Z} + \mathbf{Y}) \tag{6.106}$$

Equation (6.105) may then be rearranged,

$$\mathbf{Z} + \mathbf{Y} = E(\mathsf{A}_{11} + \mathsf{B}_{11})^{-1}\mathsf{S}(\mathbf{Z} - \mathbf{Y}) \tag{6.107}$$

and inserted into Eq. (6.106) to give

$$\mathsf{S}^{-1}(\mathsf{A}_{11} + \mathsf{B}_{11})\mathsf{S}^{-1}(\mathsf{A}_{11} - \mathsf{B}_{11})(\mathbf{Z} - \mathbf{Y}) = E^2(\mathbf{Z} - \mathbf{Y}) \tag{6.108}$$

The eigenvalues of Eq. (6.102) are thus determined by the nonhermitian eigenvalue problem given in Eq. (6.108) for E^2. If $\mathsf{A}_{11} - \mathsf{B}_{11}$ is positive definite, we can form the $(\mathsf{A}_{11} - \mathsf{B}_{11})^{1/2}$ matrix and premultiply Eq. (6.108) with $(\mathsf{A}_{11} - \mathsf{B}_{11})^{1/2}$, thereby achieving the hermitian eigenvalue problem

$$(\mathsf{A}_{11} - \mathsf{B}_{11})^{1/2}\mathsf{S}^{-1}(\mathsf{A}_{11} + \mathsf{B}_{11})\mathsf{S}^{-1}(\mathsf{A}_{11} - \mathsf{B}_{11})^{1/2}(\mathsf{A}_{11} - \mathsf{B}_{11})^{1/2}(\mathbf{Z} - \mathbf{Y}) = E^2(\mathsf{A}_{11} - \mathsf{B}_{11})^{1/2}(\mathbf{Z} - \mathbf{Y}) \tag{6.109}$$

which has eigenvalues E^2 and eigenvectors equal to $(\mathsf{A}_{11} - \mathsf{B}_{11})^{1/2}(\mathbf{Z} - \mathbf{Y})$. The eigenvalues of the nonhermitian eigenvalue problem in Eq. (6.102) can thus be determined from a hermitian eigenvalue problem of only the dimension of the particle–hole operators. When S is singular or nearly singular, it may be useful to solve Eq. (6.108) or (6.109) using the inverse eigenvalue equations with eigenvalues $1/E^2$. Equation (6.108) then becomes

$$(\mathsf{A}_{11} - \mathsf{B}_{11})^{-1}\mathsf{S}(\mathsf{A}_{11} + \mathsf{B}_{11})^{-1}\mathsf{S}(\mathbf{Z} - \mathbf{Y}) = (1/E^2)(\mathbf{Z} - \mathbf{Y}) \tag{6.110}$$

To interpret how transition moments are determined within the TDHF approximation, we continue transforming the propagator to its spectral form. We use the eigenvalues and eigenvectors of Eq. (6.108) together with Eq. (6.107) to determine the Z and Y matrices. Equation (6.102) implies that if the set $\binom{\mathsf{Z}}{\mathsf{Y}}$ are eigenvectors corresponding to the eigenvalues ω, then $\binom{\mathsf{Y}}{\mathsf{Z}}$ are eigenvectors with $-\omega$ eigenvalues. This allows us to write Eq. (6.102) in a form that displays its positive and negative eigenvalue spectrum

$$\begin{pmatrix} \mathsf{A}_{11} & \mathsf{B}_{11} \\ \mathsf{B}_{11} & \mathsf{A}_{11} \end{pmatrix}\begin{pmatrix} \mathsf{Z} & \mathsf{Y} \\ \mathsf{Y} & \mathsf{Z} \end{pmatrix} = \begin{pmatrix} \mathsf{S} & 0 \\ 0 & -\mathsf{S} \end{pmatrix}\begin{pmatrix} \mathsf{Z} & \mathsf{Y} \\ \mathsf{Y} & \mathsf{Z} \end{pmatrix}\begin{pmatrix} \omega & 0 \\ 0 & -\omega \end{pmatrix} \tag{6.111}$$

or alternatively as

$$\begin{pmatrix} E\mathsf{S} + \mathsf{A}_{11} & \mathsf{B}_{11} \\ \mathsf{B}_{11} & -E\mathsf{S} + \mathsf{A}_{11} \end{pmatrix}\begin{pmatrix} \mathsf{Z} & \mathsf{Y} \\ \mathsf{Y} & \mathsf{Z} \end{pmatrix} = \begin{pmatrix} \mathsf{S} & 0 \\ 0 & -\mathsf{S} \end{pmatrix}\begin{pmatrix} \mathsf{Z} & \mathsf{Y} \\ \mathsf{Y} & \mathsf{Z} \end{pmatrix}\begin{pmatrix} E\mathbf{1} + \omega & 0 \\ 0 & E\mathbf{1} - \omega \end{pmatrix} \tag{6.112}$$

Because of the appearance of the metric matrix $\left(\begin{smallmatrix} \mathsf{S} & 0 \\ 0 & -\mathsf{S} \end{smallmatrix}\right)$ in Eq. (6.102) the $\binom{\mathsf{Z}}{\mathsf{Y}}$ eigenvectors may be normalized according to

$$(\mathbf{Z}, \mathbf{Y})_\lambda \begin{pmatrix} \mathsf{S} & 0 \\ 0 & -\mathsf{S} \end{pmatrix}\begin{pmatrix} \mathbf{Z} \\ \mathbf{Y} \end{pmatrix}_\mu = \delta_{\lambda\mu} \tag{6.113}$$

To obtain a spectral representation of the propagator that contains a unit metric, one must transform the set of particle–hole and hole–particle operators to the representation where they give a diagonal metric with unit elements. This transformation is carried out using the excitation operators defined below:

$$\mathbf{0}^+ = \mathbf{Q}^+\mathsf{Z} + \mathbf{Q}\mathsf{Y} \tag{6.114}$$

$$\mathbf{0} = \mathbf{Q}^+\mathsf{Y}^+ + \mathbf{Q}\mathsf{Z}^+ \tag{6.115}$$

For example, it is straightforward to show using Eqs. (6.98), (6.99), and (6.113) that

$$(0^+|0^+) = (\mathsf{Z}^+, \mathsf{Y}^+)\begin{pmatrix} \mathsf{S} & 0 \\ 0 & -\mathsf{S} \end{pmatrix}\begin{pmatrix} \mathsf{Z} \\ \mathsf{Y} \end{pmatrix} = 1 \tag{6.116}$$

This condition then implies that the full metric matrix involving these new excitation operators becomes

$$\begin{pmatrix} (0^+|0^+) & (0|0^+) \\ (0^+|0) & (0|0) \end{pmatrix} = \begin{pmatrix} 1 & 0 \\ 0 & -1 \end{pmatrix} = \begin{pmatrix} \mathsf{Z}^+ & \mathsf{Y}^+ \\ \mathsf{Y}^+ & \mathsf{Z}^+ \end{pmatrix}\begin{pmatrix} \mathsf{S} & 0 \\ 0 & -\mathsf{S} \end{pmatrix}\begin{pmatrix} \mathsf{Z} & \mathsf{Y} \\ \mathsf{Y} & \mathsf{Z} \end{pmatrix} \tag{6.117}$$

The spectral form of the propagator is then obtained by taking the inverse of Eq. (6.112), premultiplying with $\left(\begin{smallmatrix} \mathsf{Z} & \mathsf{Y} \\ \mathsf{Y} & \mathsf{Z} \end{smallmatrix}\right)$, and using Eq. (6.117):

$$\begin{pmatrix} E\mathsf{S} + \mathsf{A}_{11} & \mathsf{B}_{11} \\ \mathsf{B}_{11} & -E\mathsf{S} + \mathsf{A}_{11} \end{pmatrix}^{-1} = \begin{pmatrix} \mathsf{Z} & \mathsf{Y} \\ \mathsf{Y} & \mathsf{Z} \end{pmatrix}\begin{pmatrix} E\mathsf{1} + \omega & 0 \\ 0 & E\mathsf{1} - \omega \end{pmatrix}^{-1}\begin{pmatrix} \mathsf{Z}^+ & \mathsf{Y}^+ \\ -\mathsf{Y}^+ & -\mathsf{Z}^+ \end{pmatrix} \tag{6.118}$$

Introducing Eq. (6.118) into Eq. (6.101) finally allows us to write the propagator in spectral form

$$\langle\langle \mathbf{r}; \mathbf{r} \rangle\rangle_E = \sum_\lambda |(\mathbf{r}|0_\lambda^+)|^2 \left[-\frac{1}{E - \omega_\lambda} + \frac{1}{E + \omega_\lambda} \right] \tag{6.119}$$

A comparison of Eq. (6.119) and the spectral representation of the propagator given in Eq. (6.86) shows that the pole at $E = \omega_\lambda$ corresponds to a total energy difference $E_\lambda - E_0$. The pole at $E = -\omega_\lambda$ corresponds to the same total energy difference $E_\lambda - E_0$, and the propagator therefore is an even function in the excitation energy $E_\lambda - E_0$. The pole at $E = \omega_\lambda$ has a residue of $-|(\mathbf{r}|0_\lambda^+)|^2$, which using Eq. (6.6) may be identified as $-|\langle 0|\mathbf{r}|\lambda\rangle|^2$. The pole at $E = -\omega_\lambda$ has the residue $|(\mathbf{r}|0_\lambda^+)|^2$, which is equal to $|\langle 0|\mathbf{r}|\lambda\rangle|^2$. The transition moments $\langle 0|\mathbf{r}|n\rangle$ may thus be determined from the residue at either of the poles $E = \pm\omega_\lambda$. It should be noted that the above development allows $E_\lambda - E_0$ to be either positive or negative corresponding to excitation energies from ground or excited states. However, in applications where $|0\rangle$ refers to an excited state, Eq. (6.108) must be used to determine the excitation energies, because $\mathsf{A} - \mathsf{B}$ is not then positive definite.

b. The Stability Condition

If imaginary or negative roots are encountered when solving the non-hermitian eigenvalue problem in Eq. (6.108), the RPA approximation is said to have an instability. If $\mathsf{A}_{11} - \mathsf{B}_{11}$ is positive definite, instabilities are not encountered if the matrix $(A_{11} - B_{11})^{1/2}S^{-1}(A_{11} + B_{11})S^{-1}(A_{11} - B_{11})^{1/2}$ in Eq. (6.109) is positive definite; that is, if

$$\mathbf{X}(A_{11} - B_{11})^{1/2}S^{-1}(A_{11} + B_{11})S^{-1}(A_{11} - B_{11})^{1/2}\mathbf{X}^T > 0 \quad (6.120)$$

for any vector satisfying $|\mathbf{X}| > 0$. Defining the vector

$$\mathbf{Y} = \mathbf{X}(A_{11} - B_{11})^{1/2}S^{-1} \quad (6.121)$$

we may write Eq. (6.120) as

$$\mathbf{Y}(A_{11} + B_{11})\mathbf{Y}^T > 0 \quad (6.122)$$

which says that $A_{11} + B_{11}$ has to be positive definite to ensure that Eq. (6.120) is fulfilled. Thus if $A_{11} - B_{11}$ is positive definite and $A_{11} + B_{11}$ is not, then an RPA instability will be encountered. Although it is not obvious from the previous derivation of the solution to the RPA problem, it may be shown by transforming the RPA eigenvalue problem to an equation similar to Eq. (6.108) (but with $Z + Y$ occurring as the eigenvector) that if $A_{11} + B_{11}$ is positive definite, then an RPA instability is encountered if $A_{11} - B_{11}$ is not positive definite. Hence if both $A_{11} \pm B_{11}$ are positive definite, instabilities are not encountered in the RPA approximation. If both $A_{11} \pm B_{11}$ are nonpositive definite, an explicit solution of Eq. (6.108) has to be determined before it be clear whether an instability is encountered. If $A_{11} \pm B_{11}$ are both non–positive-definite, negative excitation energies $(E_\lambda - E_0)$ are obtained in TDHF approximations. Such negative excitation energies may correspond to excitations from higher to lower "excited" states.

As was demonstrated in Chapter 2, the curvature of the energy hypersurface at a stationary point corresponding to the reference state $|0\rangle$ is governed by the same $A_{11} - B_{11}$ matrix [Eq. (2.80)] as occurs here in the TDHF. Hence, if the HF wavefunction corresponds to a local energy minimum, $A_{11} - B_{11}$ would be positive definite. In our derivation of the energy optimization conditions as given in Chapter 2, we restricted our orbital variations to involve only real variational parameters (i.e., we assumed real spin-orbitals). If we had instead examined the variations in the energy resulting from purely imaginary orbital variational parameters, the second derivative of the total energy would involve the matrix $A_{11} + B_{11}$. Hence, the conditions that $A_{11} \pm B_{11}$ be positive definite must be met if the HF reference state is to represent a local energy minimum both with respect to real and imaginary orbital variations. Therefore, imaginary excitation energies arise in RPA if one of the matrices $A_{11} \pm B_{11}$ is non–positive-definite and the other is positive definite. If negative excitation energies are obtained in the RPA approximation both $A_{11} \pm B_{11}$ are non–positive-definite and the reference state $|0\rangle$ then represents a saddle point on the energy hypersurface.

c. Connection with Coupled Hartee–Fock Theory

Having defined the TDHF problem and having shown how excitation energies and oscillator strengths are determined, we now demonstrate that the above TDHF propagator reduces, for $E = 0$, to the equation obtained in Chapter 5 for the second-order response property as expressed in the coupled Hartree–Fock (CHF) approach. For $E = 0$, the TDHF polarization propagator given in Eq. (6.101) becomes

$$\langle\langle \mathbf{r};\mathbf{r}\rangle\rangle_{E=0} = ((\mathbf{r}|\mathbf{Q}^+)(\mathbf{r}|\mathbf{Q}))\begin{pmatrix} \mathsf{A}_{11} & \mathsf{B}_{11} \\ \mathsf{B}_{11} & \mathsf{A}_{11} \end{pmatrix}^{-1}\begin{pmatrix} (\mathbf{Q}^+|\mathbf{r}) \\ (\mathbf{Q}|\mathbf{r}) \end{pmatrix} \tag{6.123}$$

Inserting unit matrices in the form

$$\mathsf{UU}^+ = 1 \tag{6.124}$$

where

$$\mathsf{U} \equiv \frac{1}{\sqrt{2}}\begin{pmatrix} 1 & -1 \\ 1 & 1 \end{pmatrix} \tag{6.125}$$

before and after the above inverse matrix, allows us to express the inverse matrix as

$$\begin{pmatrix} \mathsf{A}_{11} & \mathsf{B}_{11} \\ \mathsf{B}_{11} & \mathsf{A}_{11} \end{pmatrix}^{-1} = \mathsf{U}\begin{pmatrix} \mathsf{A}_{11} + \mathsf{B}_{11} & 0 \\ 0 & \mathsf{A}_{11} - \mathsf{B}_{11} \end{pmatrix}^{-1}\mathsf{U}^+ \tag{6.126}$$

Because the dipole operator $\mathbf{r}$ is real, the elementary definition of the super-operator scalar product given in Eq. (6.16) can be used to write

$$(\mathbf{r}|\mathbf{Q}) = -(\mathbf{r}|\mathbf{Q}^+) \tag{6.127}$$

which, together with Eq. (6.126), allows us to rewrite Eq. (6.123) as

$$\langle\langle \mathbf{r};\mathbf{r}\rangle\rangle_{E=0} = 2(\mathbf{r}|\mathbf{Q})(\mathsf{A}_{11} - \mathsf{B}_{11})^{-1}(\mathbf{Q}|\mathbf{r}) \tag{6.128}$$

Comparing this expression to that of the CHF approach [Eq. (5.16)] shows that these two ways of writing the frequency-independent polarizability are indeed identical.

d. Equivalence of Length and Velocity Oscillator Strengths

Another important and attractive feature of the TDHF approximation (and its MC extension described below) is that the oscillator strengths computed within the dipole length and dipole velocity approximations become formally equivalent, provided that a complete basis is used in the calculation. From Eq. (6.119) it is clear that the transition moments in the dipole velocity approximation would given by

$$(\mathbf{p}|0_\lambda^+) = \langle 0|[\mathbf{p}, 0_\lambda^+]|0\rangle \tag{6.129}$$

which from Eq. (6.91) is equivalent to

$$-i\langle 0|[[\mathbf{r}, H], 0_\lambda^+]|0\rangle \tag{6.130}$$

Using the matrix identity of Eq. (6.42) together with the BT theorem in the form

$$\langle 0|[[\mathbf{r}, 0_\lambda^+], H]|0\rangle = 0 \tag{6.131}$$

we can express the above as

$$(\boldsymbol{p}|0_\lambda^+) \;=\; -i\langle 0|[\mathbf{r}, [H, 0_\lambda^+]]|0\rangle = -i\boldsymbol{r}\langle 0|[\mathbf{Q}^+ + \mathbf{Q}, [H, 0_\lambda^+]]|0\rangle \tag{6.132}$$

where $\boldsymbol{r}$ denotes a row vector that contains the particle–hole matrix elements $(\boldsymbol{r})_{m\alpha}$. Equation (6.132) may be rewritten, using the definitions of 0_λ^+ [Eq. (6.114)] and the A_{11} and B_{11} matrices, as

$$(\boldsymbol{p}|0_\lambda^+) = -i(\boldsymbol{r}, \boldsymbol{r})\begin{pmatrix} \mathsf{A}_{11} & \mathsf{B}_{11} \\ \mathsf{B}_{11} & \mathsf{A}_{11} \end{pmatrix}\begin{pmatrix} \mathbf{Z} \\ \mathbf{Y} \end{pmatrix}_\lambda = -\omega_\lambda i(\boldsymbol{r}, \boldsymbol{r})\begin{pmatrix} \mathsf{S} & 0 \\ 0 & -\mathsf{S} \end{pmatrix}\begin{pmatrix} \mathbf{Z} \\ \mathbf{Y} \end{pmatrix}_\lambda \tag{6.133}$$

The last equality sign follows from the eigenvalue relation Eq. (6.102). Since

$$((\mathbf{r}|\mathbf{Q}^+)(\mathbf{r}|\mathbf{Q})) = (\boldsymbol{r}, \boldsymbol{r})\begin{pmatrix} \mathsf{S} & 0 \\ 0 & -\mathsf{S} \end{pmatrix} \tag{6.134}$$

we may finally rewrite Eq. (6.133) as

$$(\mathbf{p}|0_\lambda^+) = -i\omega_\lambda((\mathbf{r}|\mathbf{Q}^+)(\mathbf{r}|\mathbf{Q}))\begin{pmatrix} \mathbf{Z} \\ \mathbf{Y} \end{pmatrix}_\lambda = -i\omega_\lambda(\mathbf{r}|0_\lambda^+) \tag{6.135}$$

where the last step follows from the definition of the excitation operator in Eq. (6.114). Equation (6.135) states that oscillator strengths calculated in the dipole length and in the dipole velocity approximation become identical provided that the commutator relation in Eq. (6.91) is valid. Violation of the commutation relation [Eq. (6.91)] occurs when a finite basis is used in the calculation.

The TDHF approximation thus has three very characteristic features that make it especially useful as a means for calculating excitation energies and oscillator strengths. In ground-state calculations it indicates via imaginary excitation frequencies if the ground state is not stable under the type of one-electron perturbation given by the choice of A and B^+. A singlet instability is thus encountered if A and B^+ are chosen to be the dipole operator, whereas triplet instabilities are obtained if A and B^+ are chosen to be, for example, the Fermi contact Hamiltonian. Second, when the energy parameter E is set equal to zero, one obtains the same expression for the second-order properties in the TDHF method as in the CHF approach. Finally, the oscillator strengths calculated within the dipole length and the dipole velocity approximations are formally equivalent. These attractive features are unique to such an approximate theory.

3. The Multiconfigurational Extension of TDHF

In many calculations on highly correlated or open-shell molecules, it turns out that a single-configuration reference state description of $|0\rangle$ is inadequate even if optimized orbitals are used to describe $|0\rangle$. If the excitation operators of interest belong to the totally symmetric irreducible representation of the Hamiltonian's point group, the results obtained are normally better than if one attempts to calculate properties whose operators are not totally symmetric (e.g., triplet operators). It is well recognized, for example, that the singlet excitation energies for a closed-shell molecule, are described relatively well (to about 10% accuracy) within the TDHF approximation, while the description of the triplet excitation energies is very poor. In fact, triplet instabilities are often encountered when using the above TDHF method.

Approximations that go beyond the simple TDHF approximation are therefore needed. We consider two such approaches here. The second method outlined below is based on a RSPT analysis in which reference state $|0\rangle$ is expanded in powers of the residual electronic interaction [given by U in Eq. (3.35)] and the projection manifold $\{\mathbf{T}^+\}$ is chosen to be large enough to guarantee that all terms in the PP are determined consistent through second order. In light of this order analysis, it will be seen that the TDHF approximation corresponds to the approximation that is consistent through first order in the electronic repulsion. Before presenting this RSPT treatment, we address another approximation that goes beyond the single-configuration TDHF approximation. This extension, which is based upon an MCSCF description of the reference state $|0\rangle$, has the same three useful characteristics mentioned above in describing the single-configuration-based TDHF description. The multiconfigurational time-dependent Hartree–Fock (MCTDHF) approximation thus provides a formalism in which oscillator strengths in the dipole length and velocity approximation remain equivalent as one ranges continuously through (MCTDHF) from a single-configuration description (TDHF) all the way to the full-CI limit.

a. *Choice of Reference Function and Operator Manifold*

Having now motivated the consideration of more sophisticated reference states, let us develop the above-mentioned approximation in some detail. In the MCTDHF approach (Yeager and Jørgensen, 1979; Dalgaard, 1980) an MCSCF wavefunction is used as the reference state. The projection manifold $\{\mathbf{T}^+\}$ is then chosen to be the same nonredundant (see Section 2.B.7) set of orbital and configuration space excitation [Eq. (2.26)] and deexcitation operators that were used for optimizing the MCSCF reference state

$$\{\mathbf{T}^+\} = \{\mathbf{Q}^+, \mathbf{R}^+, \mathbf{Q}, \mathbf{R}\} \tag{6.136}$$

b. Hermiticity Problem

This choice of $|0\rangle$ and $\{\mathbf{T}^+\}$ then permits $\langle\langle \mathbf{r}; \mathbf{r}\rangle\rangle_E$ to be written in a form analogous to that given in Eq. (6.97). In computing the requisite matrix elements, one notices that the elements giving the coupling between orbital and configuration space operators do not obey hermiticity:

$$(\mathbf{Q}^+|\hat{H}|R_n^+) - (R_n^+|\hat{H}|\mathbf{Q}^+)^* = \langle 0|H\mathbf{Q}|n\rangle - E_0\langle 0|\mathbf{Q}|n\rangle \neq 0 \tag{6.137}$$

In the limit where one has in $|0\rangle$ an exact eigenstate

$$H|0\rangle = E_0|0\rangle \tag{6.138}$$

the last two terms in Eq. (6.137) cancel, and the matrix representative of $\hat{H}$ within the $\{\mathbf{T}^+\}$ basis consequently becomes hermitian. Therefore, we are certain that this nonhermitian aspect of the problem is an artifact (i.e., it arises because we do not have an exact $|0\rangle$). To force the matrix to be hermitian even for approximate choices of $|0\rangle$, we equate $(Q^+|H|R^+)$ with $(R^+|H|Q^+)^*$. That is, we simply require the superoperator Hamiltonian to operate on the orbital space (Q^+, Q) when the coupling elements are evaluated. This choice yields a propagator that for $E = 0$ gives the same result for second-order properties as obtained in the coupled multiconfiguration HF approach. An added advantage of this order of operations is that the oscillator strengths in the dipole length and in the dipole velocity approximations become formally equivalent. If we had chosen an alternative means of imposing hermiticity on the matrix, such would not be the case.

c. Spectral Representation of the Propagator

Inserting the projection manifold defined in Eq. (6.136) into Eq. (6.32) gives

$$\langle\langle \mathbf{r};\mathbf{r}\rangle\rangle_E = [(\mathbf{r}|\mathbf{Q}^+)(\mathbf{r}|\mathbf{R}^+)(\mathbf{r}|\mathbf{Q})(\mathbf{r}|\mathbf{R})]$$

$$\times \left(E\begin{pmatrix} \mathsf{S} & \boldsymbol{\Delta} \\ -\boldsymbol{\Delta} & -\mathsf{S} \end{pmatrix} + \begin{pmatrix} \mathsf{A} & \mathsf{B} \\ \mathsf{B} & \mathsf{A} \end{pmatrix}\right)^{-1} \begin{pmatrix} (\mathbf{Q}^+|\mathbf{r}) \\ (\mathbf{R}^+|\mathbf{r}) \\ (\mathbf{Q}|\mathbf{r}) \\ (\mathbf{R}|\mathbf{r}) \end{pmatrix} \tag{6.139}$$

where

$$\mathsf{S} = \begin{pmatrix} \langle 0|[Q, Q^+]|0\rangle & \langle 0|[Q, R^+]|0\rangle \\ \langle 0|[R, Q^+]|0\rangle & \langle 0|[R, R^+]|0\rangle \end{pmatrix} \tag{6.140}$$

$$\boldsymbol{\Delta} = \begin{pmatrix} \langle 0|[Q, Q]|0\rangle & \langle 0|[Q, R]|0\rangle \\ \langle 0|[R, Q]|0\rangle & \langle 0|[R, R]|0\rangle \end{pmatrix} \tag{6.141}$$

and $\mathsf{S}^+ = \mathsf{S}^*$ and $\boldsymbol{\Delta}^+ = -\boldsymbol{\Delta}^*$. The elements of, for example, $\langle 0|[Q, R^+]|0\rangle$ are given by

$$\langle 0|[s^+r, |n\rangle\langle 0|]|0\rangle = \langle 0|s^+r|n\rangle \tag{6.142}$$

$$\langle 0|[R_n^+, R_m]|0\rangle = \langle 0|[|n\rangle\langle 0|, |0\rangle\langle m|]|0\rangle = -\langle m|n\rangle = -\delta_{mn} \tag{6.143}$$

The A and B matrices are identical to those defined in Eqs. (2.29) and (2.30) and more explicitly written out in (2.42) and (2.44). Of course, now the reference function $|0\rangle$ is the MCSCF state; in the TDHF approximation it was the single-configuration SCF state. In Problem 5.3, the A and B matrices are evaluated for a single molecular ion, and in Problem 6.3 the data are used to perform an MCTDHF calculation on that system.

Because the metric in the MCTDHF approximation [Eqs. (6.140) and (6.141)] has a more general form than the one in the TDHF approximation [Eqs. (6.98) and (6.99)] some minor modifications are required in the procedure described in Section E.2.a. to get the propagator into spectral form.

By carrying out transformations of the MTDHF eigenvalue problem, similar to that done in Eqs. (6.102)–(6.108), we obtain the result analogous to Eq. (6.107) (Jørgensen *et al.*, 1981)

$$\mathbf{Z} + \mathbf{Y} = E(\mathsf{A} + \mathsf{B})^{-1}(\mathsf{S} - \boldsymbol{\Delta})(\mathbf{Z} - \mathbf{Y}) \tag{6.144}$$

to Eq. (6.108)

$$(\mathsf{S} - \boldsymbol{\Delta})^{-1}(\mathsf{A} + \mathsf{B})(\mathsf{S} + \boldsymbol{\Delta})^{-1}(\mathsf{A} - \mathsf{B})(\mathbf{Z} - \mathbf{Y}) = E^2(\mathbf{Z} - \mathbf{Y}) \tag{6.145}$$

and to Eq. (6.109)

$$(\mathsf{A} - \mathsf{B})^{1/2}(\mathsf{S} - \boldsymbol{\Delta})^{-1}(\mathsf{A} + \mathsf{B})(\mathsf{S} + \boldsymbol{\Delta})^{-1}(\mathsf{A} - \mathsf{B})^{1/2}(\mathsf{A} - \mathsf{B})^{1/2}(\mathbf{Z} - \mathbf{Y}) = E^2(\mathsf{A} - \mathsf{B})^{1/2}(\mathbf{Z} - \mathbf{Y}) \tag{6.146}$$

Using these equations, a spectral representation may easily be derived as was done in the single-configuration case in Eqs. (6.109)–(6.119).

d. Special Characteristics of the MCTDHF Propagator

The MCSCF reference state represents a stationary point on the energy hypersurface. If imaginary excitation energies are encountered, for example, in an MCTDHF ground-state calculation, the minimum point is not stable (does not correspond to a local minimum) under the type of (spatial or spin symmetry) one-electron perturbations described by the operators A and B^+. Also, as in the single-configuration TDHF, the frequency-independent polarizability obtained in the MCTDHF approximation becomes identical to that resulting from the multiconfiguration coupled HF approach. The proof of this equivalence follows exactly the same lines as for the single-configuration case; we refer to that proof for further details [see Eqs. (6.123)–(6.128)]. The essential points of the proof are as follows. For $E = 0$, Eq. (6.139) reduces to

$$\langle\langle \mathbf{r}; \mathbf{r} \rangle\rangle_{E=0} = [(\mathbf{r}|\mathbf{Q}^+)(\mathbf{r}|\mathbf{R}^+)(\mathbf{r}|\mathbf{Q})(\mathbf{r}|\mathbf{R})] \begin{pmatrix} \mathsf{A} & \mathsf{B} \\ \mathsf{B} & \mathsf{A} \end{pmatrix}^{-1} \begin{pmatrix} (\mathbf{Q}^+|\mathbf{r}) \\ (\mathbf{R}^+|\mathbf{r}) \\ (\mathbf{Q}|\mathbf{r}) \\ (\mathbf{R}|\mathbf{r}) \end{pmatrix} \tag{6.147}$$

By next inserting the unit matrix of Eq. (6.124) before and after the above inverse matrix and then using Eq. (6.127), we can write the frequency-independent polarizability in the form

$$\langle\langle \mathbf{r}; \mathbf{r} \rangle\rangle_{E=0} = 2[(\mathbf{r}|\mathbf{Q})(\mathbf{r}|\mathbf{R})](\mathsf{A} - \mathsf{B})^{-1} \begin{pmatrix} (\mathbf{Q}|\mathbf{r}) \\ (\mathbf{R}|\mathbf{r}) \end{pmatrix} \tag{6.148}$$

which is identical to the expression obtained in the multiconfiguration coupled HF calculation of Eq. (5.15). It has further been proven by Dalgaard (1980) that the oscillator strength calculated within the dipole length and dipole velocity approximations become identical if a complete basis is used in the MCTDHF calculation. A proof that follows lines very similar to the ones given in Section E.d for a single-configuration case has been given (Albertsen *et al.*, 1980).

The MCTDHF approximation thus has the same characteristics as the single-configuration-based TDHF approximation. We therefore have the possibility of determining approximate state vectors that, at any level of approximation, show these characteristics as the number of configurations included in the MCSCF reference state is increased from the single-configuration case through the full-CI limit. Initial calculations using the MCTDHF approximation have yielded very promising results. We now move to describing an extension of the TDHF approximation that is based on perturbation theory.

4. Rayleigh–Schrödinger Analysis

The perturbation extension of the TDHF method is obtained by developing systematic approximations to the PP that are consistent through a certain order in the perturbation (Oddershede, 1978). These approximations are based upon expanding the reference state $|0\rangle$ in powers of the residual electron–electron interaction as defined in RSPT [see Eq. (3.14)] and by choosing the projection manifolds $\{\mathbf{T}_k^+\}$ to be sufficiently "large" to ensure that the resulting matrix elements of the PP propagator are consistent through the desired order.

In the EP case, our goal was to determine the primary poles (the ionization potentials and electron affinities involving low-energy ionization of the parent molecule) through second or third order. In the analogous PP calculations, the primary poles correspond to those possessing dominant particle–hole nature; we attempt to determine these poles through a chosen order. Because the residues at a given pole contain information about the transion amplitudes for the given type of excitation, perturbation methods may also be employed to evaluate these residues through a specified order. Further, since the PP expresses the reference states' frequency-dependent polarizabilities, this response quantity may also be calculated consistent through the desired order by using a PP consistent through that order.

a. *Choice of Operator Space*

As the unperturbed Hamiltonian, we choose the same HF Hamiltonian as was employed in the above EP development, and we use a basis set of real orthonormal spin-orbitals. We develop an approximation to the PP that yields the primary excitation energies and the corresponding transition moments (and the frequency-dependent polarizability) consistent through second order in the residual electronic repulsion (Nielsen *et al.*, 1980). To determine the poles belonging to the principal excitation energies, the corresponding transition moments, and the frequency-dependent polarizability through second order, it proves sufficient to consider the truncated projection manifold

$$\{\mathbf{T}^+\} = \{\mathbf{T}_2^+;\mathbf{T}_4^+\} \tag{6.149}$$

This conclusion is by no means obvious. One must, in principle, examine the effects of $\mathbf{T}_6^+$, $\mathbf{T}_8^+$, etc. on the matrix elements $(\mathbf{T}^+|\hat{H}|\mathbf{T}^+)$, $(B^+|\mathbf{T}^+)$, $(\mathbf{T}^+|A)$, and $(\mathbf{T}^+|\mathbf{T}^+)$ to conclude that these higher operators can have no effect, through second order, on the computed poles and residues of the PP (Oddershede and Jørgensen, 1977).

b. *Pole and Residue Structure of the Propagator*

With the above choice of the operator projection manifold, the PP propagator [Eq. (6.32)] becomes

$$\langle\langle \mathbf{r};\mathbf{r}\rangle\rangle_E = [(\mathbf{r}|\mathbf{T}_2^+)(\mathbf{r}|\mathbf{T}_4^+)]\begin{pmatrix}(\mathbf{T}_2^+|E\hat{I}+\hat{H}|\mathbf{T}_2^+) & (\mathbf{T}_2^+|E\hat{I}+\hat{H}|\mathbf{T}_4^+)\\ (\mathbf{T}_4^+|E\hat{I}+\hat{H}|\mathbf{T}_2^+) & (\mathbf{T}_4^+|E\hat{I}+\hat{H}|\mathbf{T}_4^+)\end{pmatrix}^{-1}\begin{pmatrix}(\mathbf{T}_2^+|\mathbf{r})\\ (\mathbf{T}_4^+|\mathbf{r})\end{pmatrix} \tag{6.150}$$

Introducing the shorthand notation

$$\mathsf{M}_{ij} = (\mathbf{T}_i^+|E\hat{I} + \hat{H}|\mathbf{T}_j^+) \tag{6.151}$$

we may partition the inverse matrix of Eq. (6.150), as was done in the EP, to yield

$$\begin{pmatrix} \mathsf{M}_{22} & \mathsf{M}_{24} \\ \mathsf{M}_{42} & \mathsf{M}_{44} \end{pmatrix}^{-1} = \begin{pmatrix} (\mathsf{M}_{22} - \mathsf{M}_{24}\mathsf{M}_{44}^{-1}\mathsf{M}_{42})^{-1} & -(\mathsf{M}_{22} - \mathsf{M}_{24}\mathsf{M}_{44}^{-1}\mathsf{M}_{42})^{-1}\mathsf{M}_{24}\mathsf{M}_{44}^{-1} \\ (-\mathsf{M}_{44}^{-1}\mathsf{M}_{42}(\mathsf{M}_{22} & \mathsf{M}_{44}^{-1} - \mathsf{M}_{44}^{-1}\mathsf{M}_{42}(\mathsf{M}_{22} \\ -\mathsf{M}_{24}^{-1}\mathsf{M}_{44}^{-1}\mathsf{M}_{42})^{-1} & -\mathsf{M}_{24}\mathsf{M}_{44}^{-1}\mathsf{M}_{42})^{-1}\mathsf{M}_{24}\mathsf{M}_{44}^{-1} \end{pmatrix} \tag{6.152}$$

By substituting the inverse matrix into Eq. (6.150) and multiplying out the factors we obtain

$$\begin{aligned} \langle\langle \mathbf{r};\mathbf{r}\rangle\rangle_E &= [(\mathbf{r}|\mathbf{T}_2^+) - (\mathbf{r}|\mathbf{T}_4^+)\mathsf{M}_{44}^{-1}\mathsf{M}_{42}]\mathsf{P}^{-1}(E)[(\mathbf{T}_2^+|\mathbf{r}) - \mathsf{M}_{24}\mathsf{M}_{44}^{-1}(\mathbf{T}_4^+|\mathbf{r})] \\ &\quad + (\mathbf{r}|\mathbf{T}_4^+)\mathsf{M}_{44}^{-1}(\mathbf{T}_4^+|\mathbf{r}) \\ &\equiv \mathsf{W}_2(E) + \mathsf{W}_4(E) \end{aligned} \tag{6.153}$$

where

$$\mathsf{P}(E) = \mathsf{M}_{22} - \mathsf{M}_{24}\mathsf{M}_{44}^{-1}\mathsf{M}_{42} \tag{6.154}$$

The principal poles of the propagator occur at the eigenvalues of $\mathsf{P}(E)$. Therefore, to obtain these poles consistent through second order, we require $\mathsf{P}(E)$ to be determined consistent through that same order. To compute the transition amplitudes consistent through second order requires that the quantity $\mathbf{F}(E)$ defined by

$$\mathbf{F}(E) \equiv [(\mathbf{r}|\mathbf{T}_2^+) - (\mathbf{r}|\mathbf{T}_4^+)\mathsf{M}_{44}^{-1}\mathsf{M}_{42}] \tag{6.155}$$

which contains zeroth- and higher-order factors, also be evaluated consistent through second order. Finally, if the frequency-dependent polarizability is to be calculated through second order, both $\mathsf{W}_2(E)$ and the $\mathsf{W}_4(E)$ should be computed through that order. Let us now analyze in more detail which of the above matrices have to be evaluated explicitly through which order to guarantee that the above quantities are calculated consistent through second order.

c. *Second-Order Analysis of Pole Structure*

We consider initially the calculation of the excitation energies that are determined as poles of $\mathsf{P}^{-1}(E)$ [Eq. (6.154)]. After introducing the individual components $\mathbf{Q}$, $\mathbf{Q}^+$, $\mathbf{Q}^+\mathbf{Q}^+$, and $\mathbf{QQ}$, we may carry out an order analysis of the individual matrices appearing in $\mathsf{P}(E)$. We consider initially the matrix M_{22}, which is expected to be the dominant contributor to P.

$$\mathsf{M}_{22} = \begin{pmatrix} E(\mathbf{Q}^+|\mathbf{Q}^+) + (\mathbf{Q}^+|\hat{H}|\mathbf{Q}^+) & (\mathbf{Q}^+|\hat{H}|\mathbf{Q}) \\ (\mathbf{Q}|\hat{H}|\mathbf{Q}^+) & -E(\mathbf{Q}|\mathbf{Q}) + (\mathbf{Q}|\hat{H}|\mathbf{Q}) \end{pmatrix}$$
$$\equiv \begin{pmatrix} E\mathsf{S} + \mathsf{A} & \mathsf{B} \\ \mathsf{B} & -E\mathsf{S} + \mathsf{A} \end{pmatrix} \tag{6.156}$$

where

$$\mathsf{S} = (\mathbf{Q}^+|\mathbf{Q}^+) \tag{6.157}$$
$$\mathsf{A} = (\mathbf{Q}^+|\hat{H}|\mathbf{Q}^+) \tag{6.158}$$
$$\mathsf{B} = (\mathbf{Q}^+|\hat{H}|\mathbf{Q}) \tag{6.159}$$

and where we have used the fact that

$$(\mathbf{Q}^+|\mathbf{Q})_{m\gamma,n\delta} = \langle 0|[\gamma^+ m, \delta^+ n]|0\rangle = 0 \tag{6.160}$$

is identically zero through any order. This M_{22} is the same matrix that occurred in the earlier TDHF treatment of the PP except that now $|0\rangle$ represents an RSPT expansion of the reference state. As an example of how to carry out the order analysis, we consider the S and A matrices:

$$\begin{aligned} S_{m\alpha,n\beta} &= \langle 0|[\alpha^+ m, n^+\beta]|0\rangle \\ &= \langle {}^0 0|[\alpha^+ m, n^+\beta]|0^0\rangle + \langle {}^1 0|[\alpha^+ m, n^+\beta]|0^1\rangle + O(3) \\ &= (S_0)_{m\alpha,n\beta} + (S_2)_{m\alpha,n\beta} + \cdots \end{aligned} \tag{6.161}$$

where

$$(S_0)_{m\alpha,n\beta} = \delta_{mn}\delta_{\alpha\beta} \tag{6.162}$$
$$(S_2)_{m\alpha,n\beta} = \tfrac{1}{2}\delta_{mn} \sum_{\substack{pq \\ \gamma}} K^{pq}_{\alpha\gamma} K^{pq}_{\beta\gamma} - \tfrac{1}{2}\delta_{\alpha\beta} \sum_{\substack{p \\ \gamma\delta}} K^{mp}_{\gamma\delta} K^{np}_{\gamma\delta} \tag{6.163}$$

and, as in the EP analysis, the superscripts on $|0^0\rangle$, $|0^1\rangle$, etc. denote the orders of these terms. No first-order terms thus appear in S because $|0^1\rangle$ contains only doubly excited configurations. The elements of A are given by

$$\begin{aligned} A_{m\alpha,n\beta} &= \langle 0|[\alpha^+ m, [H, n^+\beta]]|0\rangle \\ &= \langle {}^0 0|[\alpha^+ m, [H_0, n^+\beta]]|0^0\rangle + \langle {}^0 0|[\alpha^+ m, [U, n^+\beta]]|0^0\rangle \\ &\quad + \langle {}^0 0|[\alpha^+ m, [U, n^+\beta]]|0^1\rangle + \langle {}^1 0|[\alpha^+ m, [H_0, n^+\beta]]|0^1\rangle + O(3) \end{aligned} \tag{6.164}$$

In the A matrix zeroth-, first-, and second-order terms appear. The A_0, A_1, and A_2 matrices are given by

$$(A_0)_{m\alpha,n\beta} = (\varepsilon_m - \varepsilon_\beta)\delta_{mn}\delta_{\alpha\beta} \tag{6.165}$$

$$(A_1)_{m\alpha,n\beta} = \langle \beta m \| n\alpha \rangle \tag{6.166}$$

$$(A_2)_{m\alpha,n\beta} = \tfrac{1}{2}\delta_{\alpha\beta} \sum_{\pi\gamma q} \langle \pi\gamma \| qn \rangle K_{\pi\gamma}^{mq} - \tfrac{1}{2}\delta_{mn} \sum_{\pi qp} \langle \beta\pi \| qp \rangle K_{\alpha\pi}^{qp} + (S_2)_{m\alpha,n\beta}(\varepsilon_n - \varepsilon_\beta) \tag{6.167}$$

The B matrix may similarly be shown to contain only first- and second-order terms $B = B_{1+2}$. The B_1 matrix is given in Problem 5.2 and

$$(B_2)_{m\alpha,n\beta} = -\sum_{q\pi} \{\langle \beta q \| \pi m \rangle K_{\pi\alpha}^{qn} + \langle \alpha q \| \pi n \rangle K_{\beta\pi}^{mq}\} - \tfrac{1}{2}\sum_{pq} \langle pq \| nm \rangle K_{\beta\alpha}^{pq} - \tfrac{1}{2}\sum_{\pi\delta} \langle \alpha\beta \| \pi\delta \rangle K_{\pi\delta}^{mn} \tag{6.168}$$

In the term $M_{24}M_{44}^{-1}M_{42}$, the M_{42} matrix contains no zeroth-order terms and since $M_{24}^T = M_{42}$, we only need keep M_{42} through first order and M_{44} through zeroth order to obtain

$$M_{24}M_{44}^{-1}M_{42} \tag{6.169}$$

through second order. The nonvanishing parts of the matrices M_{42} and M_{44} become

$$\begin{aligned}(C_1)_{nm\beta\alpha,p\gamma} &= \langle 0^0|[\alpha^+\beta^+mn,[U,p^+\gamma]]|0^0\rangle \\ &= \delta_{mp}\langle \gamma n \| \alpha\beta \rangle - \delta_{np}\langle \gamma m \| \alpha\beta \rangle \\ &\quad + \delta_{\alpha\gamma}\langle mn \| \beta p \rangle - \delta_{\beta\gamma}\langle mn \| \alpha p \rangle\end{aligned} \tag{6.170}$$

$$\begin{aligned}(D_0)_{nm\beta\alpha,pq\gamma\delta} &= E - \langle 0^0|[\alpha^+\beta^+mn,p^+q^+\gamma\delta]|0^0\rangle \\ &\quad + \langle 0^0|[\alpha^+\beta^+mn,[H_0,p^+q^+\gamma\delta]]|0^0\rangle \\ &= (E + \varepsilon_\alpha + \varepsilon_\beta - \varepsilon_m - \varepsilon_n)\delta_{np}\delta_{mq}\delta_{\beta\gamma}\delta_{\alpha\delta}\end{aligned} \tag{6.171}$$

$$(D_0)_{\alpha\beta mn,\delta\gamma qp} = (-E + \varepsilon_\alpha + \varepsilon_\beta - \varepsilon_m - \varepsilon_n)\delta_{np}\delta_{mq}\delta_{\beta\gamma}\delta_{\alpha\delta} \tag{6.172}$$

The excitation energies as computed through second order may thus be obtained as poles of

$$P^{-1}(E) = \begin{pmatrix} [ES_{0+2} + A_{0+1+2} - C_1^T D_0^{-1} C_1] & B_{1+2} \\ B_{1+2} & [-ES_{0+2} + A_{0+1+2} - C_1^T D_0^{-1} C_1] \end{pmatrix}^{-1} \tag{6.173}$$

We should also note that Eq. (6.173), when used to determine the excitation energies consistent only through first order, reduces to the inverse matrix occurring in the TDHF approximation described in Section E.2.

d. Second-Order Analysis of Transition Moments

When the excitation energies are determined through second order we might also wish to determine the corresponding transition moments consistent through the same order. This would require us to evaluate the eigenvectors of $\mathsf{P}^{-1}(E)$ and to further evaluate $\mathbf{F}(E)$ of Eq. (6.155) consistent through second order. Because the eigenvectors of $\mathsf{P}(E)$ become energy dependent, specialized techniques are required to determine the transition moments (Oddershede *et al.*, 1977).

To determine $\mathbf{F}(E)$ consistent through second order, we introduce the individual components of $\mathbf{Q}^+$, $\mathbf{Q}$, etc., which then permits the first component [see Eq. (6.155)] of $\mathbf{F}(E)$ to be expressed as

$$(\mathbf{r}|\mathbf{Q}^+) = \langle {}^0 0|[\mathbf{r}, \mathbf{Q}^+]|0^0\rangle + \langle {}^1 0|[\mathbf{r}, \mathbf{Q}^+]|0^1\rangle + \langle {}^2 0|[\mathbf{r}, \mathbf{Q}^+]|0^0\rangle + \langle {}^0 0|[\mathbf{r}, \mathbf{Q}^+]|0^2\rangle + O(3) \tag{6.174}$$

which contains zeroth- and second-order contributions. The values of $(\mathbf{r}|\mathbf{Q}^+)_0$ are given in Eq. (6.100). The only $|0^2\rangle$ terms that contribute to $(\mathbf{r}|\mathbf{Q}^+)_2$ are those which contain singly excited configurations relative to $|0^0\rangle$. The matrix $(\mathbf{r}|\mathbf{Q}^+\mathbf{Q}^+)$ has no zeroth-order elements; thus from the expression for $\mathbf{F}(E)$, it is obvious (because M_{42} is of at least first order) that only the first-order elements of $(\mathbf{r}|\mathbf{Q}^+\mathbf{Q}^+)$ can contribute. Explicit expressions for $(\mathbf{r}|\mathbf{Q}^+)_2$ and $(\mathbf{r}|\mathbf{Q}^+\mathbf{Q}^+)_1$ have been obtained (Nielsen *et al.*, 1980). The expression for $\mathbf{F}(E)$ consistent through second order may then be written as

$$\mathbf{F}(E) = \{(\mathbf{r}|\mathbf{Q}^+)_{0+2} - (\mathbf{r}|\mathbf{Q}^+\mathbf{Q}^+)_1\mathsf{D}_0^{-1}\mathsf{C}_1, (\mathbf{r}|\mathbf{Q})_{0+2} - (\mathbf{r}|\mathbf{Q}\mathbf{Q})_1\mathsf{D}_0^{-1}\mathsf{C}_1\} \tag{6.175}$$

which may be used to calculate the transition moments correct through second order. If we wish to calculate the transition moments correct only through first order, $\mathbf{F}(E)$ reduces to $\{(\mathbf{r}|\mathbf{Q}^+)_0, (\mathbf{r}|\mathbf{Q})_0\}$, which is identical to the expression for $\mathbf{F}(E)$ used in the TDHF approximation. Therefore, we again see that in the TDHF approximation, both the excitation energies and the transition moments are calculated correct through first order.

e. Frequency-Dependent Polarizability

To obtain the frequency-dependent polarizability correct through second order requires that the $\mathsf{W}_2(E)$ be evaluated consistent through second order as described previously and further that $\mathsf{W}_4(E)$ be calculated through second order. By introducing the individual compnents $\mathbf{Q}^+\mathbf{Q}^+$ and $\mathbf{Q}\mathbf{Q}$ we can reduce $\mathsf{W}_4(E)$ to

$$\mathsf{W}_4(E) = (\mathbf{r}|\mathbf{Q}^+\mathbf{Q}^+)_1\mathsf{D}_0^{-1}(\mathbf{Q}^+\mathbf{Q}^+|\mathbf{r})_1 + (\mathbf{r}|\mathbf{Q}\mathbf{Q})_1\mathsf{D}_0^{-1}(\mathbf{Q}\mathbf{Q}|\mathbf{r})_1 \quad (6.176)$$

Since $\mathsf{W}_4(E)$ contains no first-order terms, the frequency-dependent polarizability is thus determined consistent through first order in the TDHF approximation [which contains no analog to $\mathsf{W}_4(E)$].

f. Diagrammatic Analysis

We have previously shown how the results of MBPT and the perturbative analysis of the EP may be interpreted in terms of a set of diagrams. The perturbative analysis of the PP may be given a similar interpretation. We sketch in the following how the diagrammatic analysis of the PP propagator may be carried out. Initially, we limit ourselves to considering how the TDHF approximation may be understood in terms of diagrams. We consider the TDHF PP approximation in the SCF spin-orbital basis, where it reads

$$\langle\langle \mathbf{r};\mathbf{r}\rangle\rangle_E = [(\mathbf{r}|\mathbf{Q}^+)(\mathbf{r}|\mathbf{Q})]\begin{pmatrix} E\mathsf{1} + \mathsf{A}_0 + \mathsf{A}_1 & \mathsf{B}_1 \\ \mathsf{B}_1 & -E\mathsf{1} + \mathsf{A}_0 + \mathsf{A}_1 \end{pmatrix}^{-1}\begin{pmatrix} (\mathbf{Q}^+|\mathbf{r}) \\ (\mathbf{Q}|\mathbf{r}) \end{pmatrix} \quad (6.177)$$

which is identical to Eq. (6.101). The poles of the inverse matrix appearing in Eq. (6.177) may be determined from a partitioned form of the inverse matrix with $\mathbf{T}_a^+ + \mathbf{T}_b^+$ of Section C.4 equal to $\mathbf{Q}^+ + \mathbf{Q}$. The analog of Eq. (6.45) then becomes

$$\mathsf{P}^{-1}(E) = [E\mathsf{1} + \mathsf{A}_0 + \mathsf{A}_1 - \mathsf{B}_1(-E\mathsf{1} + \mathsf{A}_0 + \mathsf{A}_1)^{-1}\mathsf{B}_1]^{-1} \quad (6.178)$$

$\mathsf{P}(E)$ may be given a diagrammatic interpretation by expanding the inverse matrix as

$$\begin{aligned}(-E\mathsf{1}+\mathsf{A}_0+\mathsf{A}_1)^{-1} &= (-E\mathsf{1}+\mathsf{A}_0)^{-1} - (-E\mathsf{1}+\mathsf{A}_0)^{-1}\mathsf{A}_1(-E\mathsf{1}+\mathsf{A}_0)^{-1} \\ &\quad + (-E\mathsf{1}+\mathsf{A}_0)^{-1}\mathsf{A}_1(-E\mathsf{1}+\mathsf{A}_0)^{-1}\mathsf{A}_1(-E\mathsf{1}+\mathsf{A}_0)^{-1} + \cdots \end{aligned} \quad (6.179)$$

We then obtain

$$\begin{aligned}\mathsf{P}(E) &= E\mathsf{1} + \mathsf{A}_0 + \mathsf{A}_1 - \mathsf{B}_1(-E\mathsf{1} + \mathsf{A}_0)^{-1}\mathsf{B}_1 \\ &\quad + \mathsf{B}_1(-E\mathsf{1} + \mathsf{A}_0)^{-1}\mathsf{A}_1(-E\mathsf{1} + \mathsf{A}_0)^{-1}\mathsf{B}_1 - \cdots \end{aligned} \quad (6.180)$$

In Fig. 6.3 we have displayed the diagrammatic representation of Eq. (6.180) in terms of Hugenholtz diagrams. Using the rules in Table II of Chapter 3 for interpreting diagrams with the modifications to rule 3 similar to those discussed in Section 6.D.2.d, we may interpret the A_1 matrix as giving rise

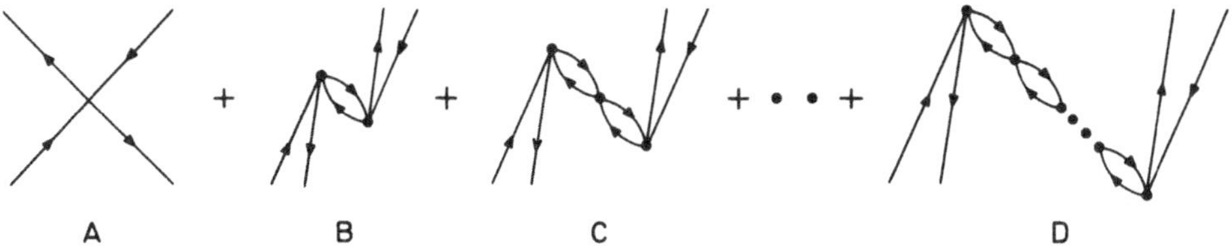

FIG. 6.3. The RPA diagram series, which can be summed to infinite order.

to diagram A in Fig. 6.3. The fourth, fifth, etc. terms in Eq. (6.180) may similarly be interpreted as giving rise to diagrams B, C, etc. in Fig. 6.3. The TDHF approximation to the PP propagator thus corresponds to summing the infinite series of diagrams represented in Fig. 6.3. We emphasize that an explicit summation of this whole series of diagrams is obtained when poles of the PP are determined as described in Section E.2.

A propagator that determines the poles consistently through second order is determined from a partitioned form of Eq. (6.173) to be

$$\mathsf{P}_2(E) = E\mathsf{1} + \mathsf{A}_{0+1+2} - \mathsf{C}_1^{\mathrm{T}}\mathsf{D}_0^{-1}\mathsf{C}_1 - \mathsf{B}_1(-E\mathsf{1} + \mathsf{A}_0)^{-1}\mathsf{B}_1 \quad (6.181)$$

All terms of order higher than two have been neglected in Eq. (6.181). In Fig. 6.4 we have displayed all the second-order PP diagrams. The A_2 matrix gives rise to diagrams A and B in Fig. 6.4, whereas the term $-\mathsf{C}^{\mathrm{T}}\mathsf{D}_0^{-1}\mathsf{C}_1$ gives diagrams C through H. The last term in Eq. (6.181) corresponds to diagram I in Fig. 6.4. This diagram is the second diagram in the above described TDHF series. We again stress that a determination of the poles of the PP that contain all diagrams through second order [Eq. (6.181)] differs from the approximation we derived in Section E.4.c, which contained all matrices of the PP through second order. A diagrammatic interpretation of this group of matrices would further contain many series of diagrams that would be summed to infinite order. One of these series would be the TDHF series given in Fig. 6.3. We do not go further into the diagrammatic

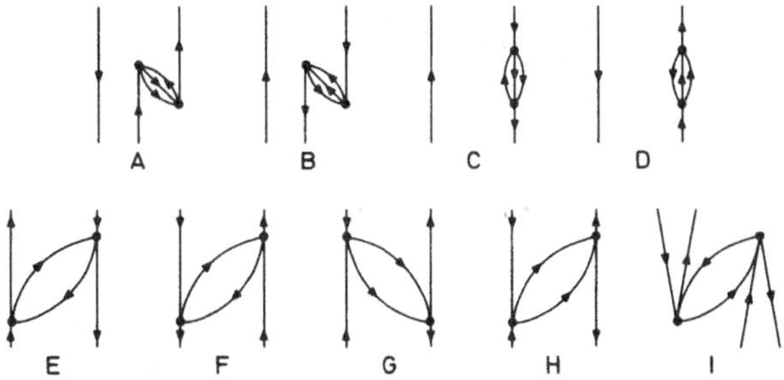

Fig. 6.4. All Hugenholtz second-order PP diagrams.

interpretation of the PP here; rather we refer the reader to the more comprehensive discussion given in the literature (Oddershede and Jørgensen, 1977).

PROBLEMS

6.1

1. Use the formulas for the second-order matrix $\mathsf{P}_2(E)$ appearing in Eq. (6.79) to express the 2×2 matrix relevant to evaluating the ionization potential and electron affinities of the minimal-basis HeH^+ problem.

2. Using the SCF orbital energies and two-electron integrals given in Problem 2.1, insert numerical values for the requisite integrals and orbital energies to express each of the elements of the 2×2 matrix $\mathsf{P}_2(E)$ as functions of E.

3. Use the approximation $(\mathsf{P}_2)_{11} = 0$ to compute the value of E at which the primary ionization potential of HeH^+ would be expected. This is done by using the Koopmans' theorem estimate in the denominators occurring in the self-energy terms and then solving for the "corrected" value of E.

4. Use the approximation $(\mathsf{P}_2)_{22} = 0$ to compute the value of E at which the primary electron affinity of HeH^+ would be expected.

5. Are the values of E found in questions 3 and 4 the only values of E that make $(\mathsf{P}_2)_{11}$ or $(\mathsf{P}_2)_{22}$ vanish?

6.2 Carry out a TDHF calculation for HeH^+, using the minimal basis data of Problem 2.1. The SCF calculation was carried out in Problem 2.1, and the matrix elements necessary for carrying out the TDHF calculation are given in Problem 5.2.

1. Determine the excitation energies and transition moments in the TDHF approximation.

2. Determine the frequency dependent polarizability tensor for $E = 0$ and for $E = 0.1$ a.u.

6.3 Carry out an MCTDHF calculation for HeH^+ that has an MCSCF reference state containing the configurations $1\sigma^2$ and $2\sigma^2$ and that uses the data of Problem 2.1. The MCSCF calculation was carried out in Problem 2.6, and most of the matrix elements necessary for carrying out the MCTDHF calculation are given in Problem 5.3.

1. Determine the excitation energies and transition moments in the MCTDHF approximation.

2. Compare the excitation energies and transition moments obtained here with the results of the full-CI calculation of Problem 5.1. Why are the two sets of results identical?

3. Determine the frequency-dependent polarizability tensor for $E = 0$ and for $E = 0.1$ a.u. in the the MCTDHF approximation. Compare the

MCTDHF polarizability with the coupled multiconfiguration HF result of Problem 5.3 and the full-CI result in Problem 5.1. Why do these three results agree?

SOLUTIONS

6.1

1. Because we have a closed-shell reference state, one can compute $(\mathsf{P}_2)_{ij}$ for i and j having $m_s = \pm 1/2$. The terms having spins $i = \alpha$, $j = \beta$, vanish since $|0\rangle$ is an eigenfunction of S_z. Let us take i and j to be α spin:

$$(\mathsf{P}_2)_{ij} = \delta_{ij}(E + \varepsilon_i) - \sum_{\substack{m<n \\ \gamma}} \frac{\langle i\gamma||mn\rangle\langle mn||j\gamma\rangle}{\varepsilon_m + \varepsilon_n - \varepsilon_\gamma + E} - \sum_{\substack{\gamma<\delta \\ m}} \frac{\langle im||\gamma\delta\rangle\langle\gamma\delta||jm\rangle}{\varepsilon_\gamma + \varepsilon_\delta - \varepsilon_m + E}$$

Because HeH^+ has only one occupied orbital, the second sum above must have $\gamma = 1\sigma\alpha$, $\delta = 1\sigma\beta$, and (because i is α spin) $m = 2\sigma\beta$. Likewise, the first sum must have $m = 2\sigma\alpha$, $n = 2\sigma\beta$, and hence $\gamma = 1\sigma\beta$. Therefore,

$$(\mathsf{P}_2)_{ij} = \delta_{ij}(E + \varepsilon_i) - \frac{\langle i1|22\rangle\langle 22|j1\rangle}{2\varepsilon_2 - \varepsilon_1 + E} - \frac{\langle i2|11\rangle\langle 11|j2\rangle}{2\varepsilon_1 - \varepsilon_2 + E}$$

2. $$(\mathsf{P}_2)_{11} = E - 1.6562 - \frac{0.0159}{1.1984 + E} - \frac{0.0382}{-3.0835 + E}$$

$$(\mathsf{P}_2)_{22} = E - 0.2289 - \frac{0.00002}{1.1984 + E} - \frac{0.0159}{-3.0835 + E}$$

$$(\mathsf{P}_2)_{12} = (\mathsf{P}_2)_{21} = + \frac{0.0006}{1.1984 + E} + \frac{0.0246}{-3.0835 + E}$$

3. $$(\mathsf{P}_2)_{11} \cong 0, \qquad E = 1.6562 + \frac{0.0159}{1.1984 + 1.6562} + \frac{0.0382}{1.6562 - 3.0835}$$

$$= 1.6350$$

This iteration process could then be continued by using this value of E to form a new $(\mathsf{P}_2)_{11}$ from which a new E could be obtained.

4. $$(\mathsf{P}_2)_{22} \cong 0, \qquad E = 0.2289 + \frac{0.00002}{1.1984 + 0.2289} + \frac{0.0159}{0.2289 - 3.0835}$$

$$= 0.2233$$

5. No. Shake-up ionizations occur near $E = \varepsilon_1 - 2\varepsilon_2$ and $E = \varepsilon_2 - 2\varepsilon_1$. These arise due to the E dependence of the denominators in the above self-energy terms.

6.2

1. The singlet excitation operator $2_\alpha^+ 1_\alpha + 2_\beta^+ 1_\beta$ gives rise to the overlap

$$S_{21,21} = (2_\alpha^+ 1_\alpha + 2_\beta^+ 1_\beta | 2_\alpha^+ 1_\alpha + 2_\beta^+ 1_\beta) = 2.00$$

because of the closed-shell reference state. For these excitation operators, the A and B matrix elements given in Problem 5.2 become

$$(A_{11})_{21,21} = 2.1464, \qquad (B_{11})_{21,21} = -0.2522$$

The nonvanishing matrix element of **r** in Problem 5.2 is

$$\langle 0|z(2_\alpha^+ 1_\alpha + 2_\alpha^+ 1_\alpha)|0\rangle = 2\langle 2|z|1\rangle = 1.0884$$

The TDHF excitation energy obtained from Eq. (6.109) is

$$E_1 = 1.0657$$

and the corresponding eigenvector is

$$Z = 0.7083, \qquad Y = 0.0418, \qquad (z|0_{12}^+) = 1.0884(0.7083 - 0.0418) = 0.7255$$

2. The nonvanishing components of the polarizability tensor are

$$\langle\langle z; z\rangle\rangle_{E=0} = 0.9878 = \frac{2|(z|0_{12}^+)|^2}{E_1}$$

$$\langle\langle z; z\rangle\rangle_{E=0.1} = 0.9965 = 2\frac{|(z|0_{12}^+)|^2 E_1}{E_1^2 - (0.1)^2}$$

6.3

1. From Solution 5.3 we can form the elements of the 2×2 S matrix

$$S_{21,21} = (2_\alpha^+ 1_\alpha + 2_\beta^+ 1_\beta | 2_\alpha^+ 1_\alpha + 2_\beta^+ 1_\beta) = \langle 0|1_\alpha^+ 1_\alpha + 1_\beta^+ 1_\beta - 2_\alpha^+ 2_\alpha - 2_\beta^+ 2_\beta|0\rangle$$

Using the density matrices of Solution 5.3, we find

$$S_{21,21} = 2(0.9968) - 2(0.0033) = 1.9870$$
$$S_{21,|1\rangle} = (2_\alpha^+ 1_\alpha + 2_\beta^+ 1_\beta|\, |n\rangle\langle 0|) = \langle 0|(1_\alpha^+ 2_\alpha + 1_\beta^+ 2_\beta)|1\rangle\langle 0|0\rangle = 0$$
$$S_{|1\rangle,|1\rangle} = (|1\rangle\langle 0|\,|\,|1\rangle\langle 0|) = 1.0000$$

$$\mathsf{A} - \mathsf{B} = \begin{pmatrix} 2.4251 & 0.4248 \\ 0.4248 & 2.2643 \end{pmatrix}$$

$$(\mathsf{A} - \mathsf{B})^{1/2} = \begin{pmatrix} 1.5510 & 0.1393 \\ 0.1393 & 1.4983 \end{pmatrix}, \qquad (\mathsf{A} - \mathsf{B})^{-1/2} = \begin{pmatrix} 0.6501 & -0.0604 \\ -0.0604 & 0.6731 \end{pmatrix}$$

$$\mathsf{S} = \begin{pmatrix} 1.9870 & 0.0 \\ 0.0 & 1.0000 \end{pmatrix}$$

From Eq. (6.109) we know that we need $(\mathsf{A}-\mathsf{B})^{1/2}\mathsf{S}^{-1}(\mathsf{A}+\mathsf{B})\mathsf{S}^{-1}(\mathsf{A}-\mathsf{B})^{1/2}$ to find the E^2 eigenvalues:

$$(\mathsf{A}-\mathsf{B})^{1/2}\mathsf{S}^{-1}=\begin{pmatrix}0.7806 & 0.1393\\ 0.0701 & 1.4983\end{pmatrix},\qquad \mathsf{S}^{-1}(\mathsf{A}-\mathsf{B})^{1/2}=\begin{pmatrix}0.7806 & 0.0701\\ 0.1393 & 1.4983\end{pmatrix}$$

Then

$$(\mathsf{A}-\mathsf{B})^{1/2}\mathsf{S}^{-1}(\mathsf{A}+\mathsf{B})\mathsf{S}^{-1}(\mathsf{A}-\mathsf{B})^{1/2}=\begin{pmatrix}1.2998 & 1.0247\\ 1.0247 & 5.1724\end{pmatrix}$$

The two eigenvalues are $E^2=1.0454$ and $E^2=5.4266$, and the corresponding eigenvectors are $(0.9705, -0.2410)$ and $(0.2410, 0.9705)$. The excitation energies are $E=1.0225$ and 2.3295; $(\mathbf{Z}-\mathbf{Y})$ is obtained for each state, according to Eq. (6.109) as

$$(\mathbf{Z}-\mathbf{Y})_1=(\mathsf{A}-\mathsf{B})^{-1/2}\begin{pmatrix}0.9705\\ -0.2410\end{pmatrix}=\begin{pmatrix}0.6455\\ -0.2208\end{pmatrix}$$

$$(\mathbf{Z}-\mathbf{Y})_2=(\mathsf{A}-\mathsf{B})^{-1/2}\begin{pmatrix}0.2410\\ 0.9705\end{pmatrix}=\begin{pmatrix}0.0981\\ 0.6386\end{pmatrix}$$

The $(\mathbf{Z}+\mathbf{Y})$ for each state can then be obtained from

$$(\mathbf{Z}+\mathbf{Y})=E(\mathsf{A}+\mathsf{B})^{-1}\mathsf{S}(\mathbf{Z}-\mathbf{Y})$$

or from

$$E^{-1}\mathsf{S}^{-1}(\mathsf{A}-\mathsf{B})(\mathbf{Z}-\mathbf{Y})$$

to yield

$$(\mathbf{Z}+\mathbf{Y})_1=\begin{pmatrix}0.7242\\ -0.2209\end{pmatrix},\qquad (\mathbf{Z}+\mathbf{Y})_2=\begin{pmatrix}0.1100\\ 0.6388\end{pmatrix}$$

Solving for $\mathbf{Z}$ and $\mathbf{Y}$ for each state and then renormalizing $(\mathbf{Z},\mathbf{Y})$ for each state such that

$$(\mathbf{Z}\mathbf{Y})\begin{pmatrix}\mathsf{S} & 0\\ 0 & -\mathsf{S}\end{pmatrix}\begin{pmatrix}\mathbf{Z}\\ \mathbf{Y}\end{pmatrix}=1=\mathbf{ZSZ}-\mathbf{YSY}$$

we obtain

$$\mathbf{Z}_1=\begin{pmatrix}0.6926\\ -0.2233\end{pmatrix},\qquad \mathbf{Y}_1=\begin{pmatrix}0.0397\\ -0.0001\end{pmatrix}$$

$$\mathbf{Z}_2=\begin{pmatrix}0.1596\\ 0.9747\end{pmatrix},\qquad \mathbf{Y}_2=\begin{pmatrix}0.0092\\ 0.0002\end{pmatrix}$$

The desired transition moments are given by $(\mathbf{r}|0^+)$, with

$$0^+ = \mathbf{Z}(Q^+_{21,}|1\rangle\langle 0|) + \mathbf{Y}(Q_{21,}|0\rangle\langle 1|)$$

The data of Solution 5.3 tell us that

$$(\mathbf{r}|Q^+_{21}) = 1.1076 = -(\mathbf{r}|Q_{21})$$
$$(\mathbf{r}||n\rangle\langle 0|) = -0.1551 = -(\mathbf{r}||0\rangle\langle n|)$$

and so

$$(\mathbf{r}|0^+_1) = (0.6926, -0.2233)\begin{pmatrix} 1.1076 \\ -0.1551 \end{pmatrix} - (0.0397, -0.0001)\begin{pmatrix} 1.1076 \\ -0.1551 \end{pmatrix}$$
$$= 0.7578$$

Likewise,

$$(\mathbf{r}|0^+_2) = 0.0146$$

2. The full-CI calculation gave excitation energies of 1.0225 and 2.3295, which is exactly what we get here. The CI transition moments are 0.7578 and 0.0144, which are almost identical to ours. The MCSCF reference state is identical to the full-CI wavefunction even though it contains only the $1\sigma^2$ and $2\sigma^2$ configuration. This is true because the orbitals used in the MCSCF wave function are optimized orbitals. The projection manifold operating on $|0\rangle$ then yields two more linearly independent functions, which, taken together with $|0\rangle$, form a three-dimensional space capable of describing the results of the full 3×3 CI problem. We thus have both the exact reference state and a complete projection manifold $\{T^+\}$, and the MCTDHF calculation therefore is able to reproduce the full-CI result of Problem 5.1.

3. $\alpha_{zz} = 2 \sum_{j=1}^{2} |(z|0^+_j)|^2 (E_j^2 - E^2)^{-1} E_j$

$$E = 0.0, \qquad \alpha_{zz} = 2\left[\frac{0.7578^2}{1.0225} + \frac{0.0146^2}{2.3295}\right] = 1.1234$$

$$E = 0.1, \qquad \alpha_{zz} = 2\left[\frac{0.7578^2(1.0225)}{1.0454 - 0.01} + \frac{0.0146^2(2.3295)}{5.4266 - 0.01}\right] = 1.1344$$

All three calculations have the potential of giving the full-CI result as discussed in question 2.

References

Albertsen, P., Jørgensen, P., and Yeager, D. L. (1980). *Int. J. Quantum Chem.* **S14**, 289.
Banerjee, A., Shepard, R., and Simons, J. (1978). *Int. J. Quantum Chem.* **S12**, 389.
Cederbaum, L. (1973). *Theor. Chim. Acta* **31**, 329.

Dalgaard, E. (1979). *Int. J. Quantum Chem.* **15**, 169.
Dalgaard, E. (1980). *J. Chem. Phys.* **72**, 816.
Doll, J. D., and Reinhardt, W. P. (1972). *J. Chem. Phys.* **57**, 1169.
Jørgensen, P. (1975). *Annu. Rev. Phys. Chem.* **26**, 359.
Jørgensen, P., and Simons, J. (1975). *J. Chem. Phys.* **63**, 5302.
Jørgensen, P., Olsen, J., and Yeager, D. L. (1981). *J. Chem. Phys.* (in press).
Linderberg, J., and Öhrn, Y. (1973). "Propagators in Quantum Chemistry." Academic Press, New York.
Linderberg, J., and Öhrn, Y. (1977). *Int. J. Quantum Chem.* **12**, 161.
Manne, R. (1977). *Chem. Phys. Lett.* **45**, 470.
Mattuck, R. D. (1967). "A Guide to Feynman Diagrams in the Many-Body Problem." McGraw-Hill, London.
Nielsen, E. S., Jørgensen, P., and Oddershede, J. (1980). *J. Chem. Phys.* **73**, 6238.
Oddershede, J. (1978). *Adv. Quantum Chem.* **11**, 275.
Oddershede, J., and Jørgensen, P. (1977). *J. Chem. Phys.* **66**, 1541.
Oddershede, J., Jørgensen, P., and Beebe, N. H. F. (1977). *Int. J. Quantum Chem.* **12**, 655.
Pickup, B. T., and Goscinski, O. (1973). *Mol. Phys.* **26**, 1013.
Purvis, G. D., and Öhrn, Y. (1974). *J. Chem. Phys.* **60**, 4043.
Redmon, L. T., Purvis, G., and Öhrn, Y. (1976). *J. Chem. Phys.* **64**, 1752.
Schaefer, H. F., and Miller, W. H., eds. (1977). "Modern Theoretical Chemistry," Vol 3. Plenum, New York.
Simons, J. (1976). *J. Chem. Phys.* **64**, 4541.
Simons, J. (1977). *Annu. Rev. Phys. Chem.* **28**, 1.
Simons, J., and Smith, W. D. (1973), *J. Chem. Phys.* **58**, 4899.
Von Niessen, W., Cederbaum, L. S., and Domcke, W. (1979). *In* "Excited States in Quantum Chemistry" (C. A. Nicolaides and D. R. Beck, eds.), Reidel, Dordrecht.
Yeager, D. L., and Jørgensen, P. (1979). *Chem. Phys. Lett.* **65**, 77.

Index

M

N

O

P

Q

R

S

T

U

V

Printed in France by Amazon
Brétigny-sur-Orge, FR

51786482R00105